PEPPERMINT

Novel Strategies for Performance Improvement

PEPPERMINT

Novel Strategies for Performance Improvement

Bilal Ahmad, PhD
Tariq Ahmad Dar, PhD
Khalid Rehman Hakeem, PhD
Editors

AAP | APPLE ACADEMIC PRESS

First edition published 2027

Apple Academic Press Inc.
1265 Goldenrod Circle, NE,
Palm Bay, FL 32905 USA

760 Laurentian Drive, Unit 19,
Burlington, ON L7N 0A4, Canada

CRC Press
2385 NW Executive Center Drive,
Suite 320, Boca Raton FL 33431

4 Park Square, Milton Park,
Abingdon, Oxon, OX14 4RN UK

Library and Archives Canada Cataloguing in Publication

Title: Peppermint : novel strategies for performance improvement / Bilal Ahmad, PhD, Tariq Ahmad Dar, PhD, Khalid Rehman Hakeem, PhD, editors.
Other titles: Peppermint (Palm Bay, Fla.)
Names: Ahmad, Bilal (Assistant professor), editor. | Dar, Tariq Ahmad, editor. | Hakeem, Khalid Rehman, editor
Description: First edition. | Includes bibliographical references and index.
Identifiers: Canadiana (print) 20260140570 | Canadiana (ebook) 20260140651 | ISBN 9781779641540 (hardcover) | ISBN 9781779641557 (ebook)
Subjects: LCSH: Tourism—Information technology—Management. | LCSH: Tourism—Data processing—Management. | LCSH: Tourism—Technological innovations—Management.
Classification: LCC G156.5.I5 S63 2026 | DDC 338.4/7910285—dc23

Library of Congress Cataloging-in-Publication Data

Names: Ahmad, Bilal, editor | Dar, Tariq Ahmad editor | Hakeem, Khalid Rehman editor

Title: Peppermint : novel strategies for performance improvement / Bilal Ahmad, Tariq Ahmad Dar, Khalid Rehman Hakeem, editors.

Description: irst edition. | Palm Bay, FL : Apple Academic Press, 2026. | Includes bibliographical references and index.

Identifiers: LCCN 2026009144 (print) | LCCN 2026009145 (ebook) | ISBN 9781779641540 hardback | ISBN 9781779641557 ebook

Subjects: LCSH: Peppermint--Yields | Crop improvement

Classification: LCC SB295.P4 P47 2026 (print) | LCC SB295.P4 (ebook)

LC record available at https://lccn.loc.gov/2026009144

LC ebook record available at https://lccn.loc.gov/2026009145

ISBN: 978-1-77964-154-0 (hbk)
ISBN: 978-1-77964-155-7 (ebk)

DOI: 10.1201/9781779641557

Dedication

This book is dedicated to our beloved parents.

"Your Lord had decreed, that you worship none save Him, and (that you show) kindness to parents. If one of them or both attain old age with you, say not 'Fie' unto them nor repulse them, but speak unto them a gracious word. And lower unto them the wing of submission through mercy and say: My Lord! Have mercy on them both, as they did care for me when I was young."

[Al-Qur'an 17:23-24]

About the Editors

Bilal Ahmad, PhD, is currently working as Assistant Professor in the Higher Education Department of Jammu and Kashmir, India. After completing his master's in Botany from Aligarh Muslim University, Aligarh, Uttar Pradesh, India, Dr. Ahmad pursued his PhD in Botany in the field of Advanced Plant Physiology until 2018. Dr. Ahmad has studied the influence of various plant growth promoters, radiation-processed polysaccharides, and nanoparticles on various medicinal plants under normal and perturbed soil environments, particularly heavy metal stress. Dr. Ahmad has been associated with teaching various disciplines of botany, notably plant physiology and biochemistry, cell and molecular biology, plant taxonomy and plant ecology, since 2018. Dr. Ahmad has presented his research at a number of national and international conferences and has received several awards. He has authored 36 research publications in high-impact factor journals and edited volumes of different books with international publishers and published two books with international publishers. In addition, Dr. Ahmad has also reviewed various articles for high-impact factor journals of international repute.

Tariq Ahmad Dar, PhD, has 20 publications to his credit. He has published his research findings in journals of international repute, such as *Environmental and Experimental Botany, Industrial Crops and Products, Carbohydrate Polymers, Journal of Applied Research on Medicinal and Aromatic Plants, Journal of Radiation Research and Applied Sciences, Plant Physiology Reports, Journal of Essential Oil Research, Frontiers in Plant Science, Journal of Plant Growth Regulation,* etc. Dr. Dar has presented his papers at national and international conferences and has attended about 11 national and international conferences. In addition, he is a life member of the Indian Botanical Society and Indian Science Congress Association. He has guided the dissertations of 10 MSc students. He is actively involved in the aspects of enhancing growth, yield, and active constituents of medicinal and aromatic plants using plant growth regulators such as jasmonates, gibberellins, and depolymerized polysaccharides. Currently Dr. Dar is focusing on the use of marine polysaccharides as plant growth regulators and increasing the yield of active constituents of medicinal plants. The use of radiation-processed depolymerized polysaccharides, like sodium alginate, chitosan, carrageenan to alleviate heavy metal stress and different abiotic stress conditions like salinity, drought, chilling etc., are his focus areas. Dr. Dar has teaching experience of eight years at undergraduate level and two and a half years at postgraduate level. He earned his MSc in Botany from the P.G. Department of Botany, University of Kashmir, Hazratbal, Srinagar, in 2010 and PhD.in Botany from Aligarh Muslim University, Aligarh, India in 2015.

Khalid Rehman Hakeem, PhD, is Professor at King Abdulaziz University, Jeddah, Saudi Arabia. After completing his doctorate (botany; specialization in plant eco-physiology and molecular biology) from Jamia Hamdard, New Delhi, India, he worked as a lecturer at the University of Kashmir, Srinagar, India. At Universiti Putra Malaysia, Selangor, Malaysia, he was a Postdoctorate Fellow and Fellow Researcher (Associate Professor) for several years. Dr. Hakeem has more than 16 years of teaching and research experience in plant eco-physiology, biotechnology and molecular biology, medicinal plant research, plant–microbe–soil interactions, as well as in environmental studies. He is the recipient of several fellowships at both national and international levels. He has served as a visiting scientist at Jinan University, Guangzhou, China. Currently, he is involved with a number of international research projects with different government organizations. To date, Dr. Hakeem has authored and edited more than 95 books with international publishers. He also has to his credit more than 185 research publications in peer-reviewed international journals and 80 book chapters in edited volumes with international publishers.

At present, Dr. Hakeem serves as an editorial board member and reviewer for several high-impact international scientific journals. He is included in the advisory board of Cambridge Scholars Publishing, UK. He is also a fellow of the Plantae Group of the American Society of Plant Biologists, member of the World Academy of Sciences, member of the International Society for Development and Sustainability, Japan, and member of the Asian Federation of Biotechnology, Korea. Dr. Hakeem has been listed in *Marquis Who's Who in the World*, since 2014–2020. Currently, Dr. Hakeem is engaged in studying the plant processes at eco-physiological as well as molecular levels.

Contents

Contributors

Mahendra Laxman Ahire
Department of Botany Yashavantrao Chavan Institute of Science Satara, Maharashtra, India

Bilal Ahmad
Department of Botany, Government Degree college Khansahib Budgam, University of Kashmir

Khan Bilal Mukhtar Ahmed
Faculty of Agricultural Sciences, Aligarh Muslim University, Aligarh, Uttar Pradesh, India

Mehtap Aydin
Department of Molecular Biology and Genetics, Faculty of Engineering and Natural Sciences, Istanbul Sabahattin Zaim University, Istanbul, Türkiye

Mukul Machhindra Barwant
Department of Botany Yashavantrao Chavan Institute of Science Satara, Maharashtra, India

Suhail Bashir
Department of Botany, Government Degree College Sogam, Kupwara, University of Kashmir, India

Sagar Bawake
Department of Botany, Ahmednagar College Ahmednagar, Maharashtra, India

Pourya Bazyar
Hamburg University of Applied Science, Department of Mechanical Engineering and Production Management, Hamburg, Germany

Abdul Waheed Chaudhry
M. A. Kazi Institute of Chemistry, University of Sindh, Jamshoro, Pakistan

Dharini Chittaragi
Department of Plantation, Spices, Medicinal and Aromatic Crops, Kittur Rani Channamma College of Horticulture, Arabhavi, University of Horticultural Sciences, Bagalkot, Karnataka, India

Sailee Chowdhury
BCDA College of Pharmacy and Technology, Hidaypur, Barasat, Kolkata, West Bengal, India

Richard Cimler
Centre of Advanced Technologies, Faculty of Science, University of Hradec Kralove, Rokitanskeho 62, Hradec Kralove, Czech Republic

Tariq Ahmad Dar
Department of Botany, Sri Pratap College, Srinagar, Cluster University Srinagar, India

Rajni Dhalaria
Department of Biosciences, Minerva P.G. College of Arts, Science and Commerce, Indora, Himachal Pradesh, India

Nitin Tanaji Gore
Department of Botany Yashavantrao Chavan Institute of Science Satara, Maharashtra, India

Shivani Guleria
Department of Biotechnology, TIFAC-Centre of Relevance and Excellence in Agro and Industrial Biotechnology (CORE), Thapar Institute of Engineering and Technology, Patiala, India

Mohd Gulfishan
School of Agriculture & Technology, Maya Devi University, Selaqui, Dehradun, Uttarakhand

Mudasir Irfan Dar
Department of Botany, Government Degree College for Women, Kupwara, University of Kashmir, India

Umaima Ismail
Department of Pharmacognosy, Faculty of Pharmacy, University of Sindh, Jamshoro, Pakistan

Muhammad Ismail
Department of Chemistry, Kohat University of Science & Technology, Kohat, Khyber Pakhtunkhwa, Pakistan; Government Degree College Essak Khel, Lakki Marwat, Khyber Pakhtunkhwa, Pakistan

Ajmat Jahan
Department of Botany, Aligarh Muslim University, Aligarh, Uttar Pradesh, India

Koyel Kar
BCDA College of Pharmacy and Technology, Hidaypur, Barasat, Kolkata, West Bengal, India

Dipanjan Karati
School of Pharmacy, Techno India University, West Bengal, Kolkata, India

M. Masroor A. Khan
Advanced Plant Physiology Section, Department of Botany, Aligarh Muslim University, Aligarh, Uttar Pradesh, India

Neetika Kimta
School of Biological and Environmental Sciences, Shoolini University of Biotechnology and Management Sciences, Solan, Himachal Pradesh, India

Amna Kousar
Department of Biotechnology, University of Sialkot, Sialkot, Pakistan

Kamil Kuča
Centre for Basic and Applied Research, University of Hradec Kralove, Hradec Kralove, Czech Republic

Harsh Kumar
Centre for Advanced Innovation Technologies, VSB-Technical University of Ostrava, Ostrava-Poruba, Czech Republic

Subhabrota Majumdar
Calcutta Institute of Pharmaceutical Technology & Allied Health Sciences, Uluberia, Howrah, West Bengal, India

Sivakumar Manickam
Chemical and Energy Engineering, Faculty of Engineering, Universiti Teknologi Brunei, Bandar Seri Begawan, BE1410, Brunei

Rana Mazumder
Bengal College of Pharmaceutical Technology, Dubrajpur, Birbhum, West Bengal, India

Vidya Mhaske
Department of Botany, Ahmednagar College Ahmednagar, Maharashtra, India

Filza Mustafa
Atta-ur-Rahman School of Applied Biosciences, National University of Sciences and Technology, Islamabad, Pakistan

Nayab Nadeem
Department of Chemistry, University of Sialkot, Sialkot, Pakistan

Jafar Fathi Qarachal
Tarbiat Modares University, Plant Pathology Department, Tehran, Iran

Lokesh S.
Assistant Professor, Department of Agricultural Economics, Post-Graduate College of Agriculture, RPCAU, Pusa, Bihar

Asfia Shabbir
Advanced Plant Physiology Section, Department of Botany, Aligarh Muslim University, Aligarh, Uttar Pradesh, India

Ehsan Sheidaee
Tarbiat Modares University, Biosystems Engineering Department, Tehran, Iran

Hadia Shoaib
National Centre of Excellence in Analytical Chemistry, University of Sindh, Jamshoro, Pakistan

Ahmed Raza Sidhu
M. A. Kazi Institute of Chemistry, University of Sindh, Jamshoro, Pakistan

Balwant Singh
Department of Botany K. S. Saket PG College, Dr. Ram Manohar Lohia Avadh University Ayodhya, Uttar Pradesh, India

R. Suresh
Directorate of Agribusiness Development, TNAU, Coimbatore, Tamil Nadu, India

Ajar Nath Yadav
Department of Genetics, Plant Breeding and Biotechnology, Dr. Khem Singh Gill Akal College of Agriculture, Eternal University, Sirmour, Himachal Pradesh, India

Mehtab Yasmeen
Telangana Mahila Viswavidyalayam, Hyderabad, Telangana, India

Preface

Interest and support for the efficient development and production of medicinal plants and the bioactive compounds with medicinal properties obtained from them is increasing at a swift rate in all parts of the globe. This is due to an escalating recognition specified to the position of medicinal plants in the provision of ethnically pertinent and affordable health care. Moreover, it is vital for creating sustainable livelihoods in addition for the conservation of biodiversity. The present era is witnessing escalated demands of such pharmaceutically imperative bioactive compounds in view of their capacity to decrease the continuously growing health risks due to constantly changing environment. World Health Organization reports have revealed the reliance of 80% of the world's developing countries' population on herbal medicines to serve their primary health care needs. Furthermore, use of traditional medicine is quite prevalent even in the developed countries. Medicinal plants serve as the treasure of traditional medicine throughout the world, whether in developed or developing countries, owing to the potential recognition of these natural products having few side effects, are non-narcotic, are easily available at inexpensive prices, and sometimes are the only source of health-care available to the poor.

Mentha piperita L., is one of the most extensively consumed medicinal herbs globally and serves diverse usage in industrial, culinary, as well as medicinal fields. Different studies conducted on peppermint during the last decade have signified the insecticidal, herbicidal, cosmeceutical, pharmaceutical, as well as ethnomedicinal properties. Natural antioxidants have grabbed considerable interest for their usage in the food and animal industries. The antioxidant potential of peppermint is attributed to the significant content of phenols aqueous extracts and essential oil, which adds to the medicinal significance of the plant and making it an interim substitute treatment for irritable bowel syndrome.

The medicinal significance of many plants is attributed to their volatile essential oils, which possess antimicrobial, anti-inflammatory, antiviral, and antioxidant properties. Despite substantial advances in modern pharmacology, the use of medicinal plant products as effective alternative medicine has increased significantly in recent years.

India holds around 60% share of the total world mint oil production. The content of essential oils in peppermint seldom exceeds 1%, and the abiotic stressors of nature further curtail crop yield, adding to the distress of farmers by callously affecting their economy.

A variety of abiotic environmental constraints, including salinity, metal/metalloid, heat, cold, UV, light etc., act as major limiting factors for peppermint productivity. Abiotic stress is currently being realized as one of ubiquitous and potential threats to human existence, resulting in the overwhelming consequences to our health as well as agricultural systems. Decreased crop yield and soil fertility, along with the accumulation of harmful elements in food chains are jeopardizing human survival, a detrimental consequence of abiotic stress to humanity, leading to severe health and ecological consequences.

To overcome such overexploitation, different types of agricultural strategies have been employed among which application of diverse plant growth promoters (PGPs) constitutes an efficient one. The different PGPs include natural phytohormones and their synthetic derivatives, radiation-processed polysaccharides, and nanoparticles. Compared to phytohormones, radiation-processed polysaccharides are cheap and eco-friendly. Additionally, nanoparticles have also been frequently tested during the last decade to mitigate the ill effects of diverse abiotic constraints and to enhance the productivity.

This book highlights the key advancements in peppermint research in the past two decades, including cultivation practices, medicinal significance of the plant, composition of the essential oil, effect of abiotic stressors, and strategies to enhance the productivity of the plant under normal and perturbed environmental conditions. The book will prove an ideal information source for a wide scientific audience including students and research scholars.

Lastly, we extend our heartfelt appreciation to Apple Academic Press for their extraordinary support helping us to achieve this success.

CHAPTER 1

Introduction and Brief Overview of Peppermint and Its Essential Oil

HADIA SHOAIB,[1*] AHMED RAZA SIDHU,[2] UMAIMA ISMAIL,[3] MUHAMMAD ISMAIL,[4,5] and ABDUL WAHEED CHAUDHRY[2]

[1]*National Centre of Excellence in Analytical Chemistry, University of Sindh, Jamshoro, Pakistan*

[2]*Dr M. A. Kazi Institute of Chemistry, University of Sindh, Jamshoro, Pakistan*

[3]*Department of Pharmacognosy, Faculty of Pharmacy, University of Sindh, Jamshoro, Pakistan*

[4]*Department of Chemistry, Kohat University of Science & Technology, Kohat, Khyber Pakhtunkhwa, Pakistan*

[5]*Government Degree College Essak Khel, Lakki Marwat, Khyber Pakhtunkhwa, Pakistan*

*Corresponding author

ABSTRACT

Medicinal plants have long been integral to traditional therapies, valued for their diverse pharmacological properties. Various mint species have been used by humans for more than 2000 years. Cultivated widely in subtropical regions, including Asia, Europe, the United States, and the Mediterranean, peppermint is economically important due to its characteristic aroma and broad applications in the pharmaceutical, food, and cosmetic industries. Historical records indicate its use in traditional medicine for several centuries, with peppermint oil recognized as one of the most extensively

Peppermint: Novel Strategies for Performance Improvement. Bilal Amhad, Tariq Ahmad Dar, and Khalid Rehman Hakeem (Eds.)

DOI: 10.1201/9781779641557-1

employed essential oils. Its pharmacological relevance is attributed to a rich phytochemical profile dominated by menthol, menthone, neomenthol, and iso-menthone. These bioactive secondary metabolites confer a wide spectrum of biological activities, including antimicrobial, anti-inflammatory, antiviral, neuroprotective, anticancer, antioxidant, and immunomodulatory effects. Increasing scientific evidence highlights peppermint oil's therapeutic potential in supporting vital systems such as the gastrointestinal, respiratory, hepatic, renal, neurological, and dermal systems. This chapter provides an overview of peppermint and its essential oil, focusing on its historical context, botanical characteristics, phytochemical composition, extraction techniques, and broad medicinal and industrial applications.

1.1 INTRODUCTION

Natural products have demonstrated their immense value as essential resources for the development of drugs. The exploration and replenishment of medically beneficial plant-based compounds continue to garner growing interest. Currently, about 80% of the world's population employs roughly 20,000 distinct plant species in diverse traditional medicinal practices, offering alternative therapeutic options for health and safety [1]. Medicinal plants have been harnessed for their healing properties for centuries in traditional medicine. Mint varieties, including peppermint, have been utilized by humans for over two millennia, with peppermint alone boasting a history of more than 250 years of use [2]. Medicinal herbs are characterized by their capacity to prevent various diseases, and their therapeutic attributes primarily originate from the presence of phytochemicals, which are harnessed for the treatment of a diverse array of health issues [3]. A widely recognized medicinal herb is peppermint (*Mentha piperita* L.), commonly referred to as "pudina." Primarily grown in hilly regions, especially those with cold climates, this herb has garnered worldwide recognition for its calming and cooling characteristics. It is prominently featured in Ayurveda, where it is utilized to harmonize the "vata dosha" and alleviate the "pitta" and "kapha doshas" [4, 5].

Menthol is a naturally occurring substance found in peppermint, existing in the form of colorless crystals or powder. It holds a crucial role in the muscle-relaxing properties of peppermint. It has been noted for its ability to stimulate the flow of bile, relax the esophageal sphincter, promote belching, and possess antibacterial properties. Additionally, peppermint serves as a plentiful source of polyphenolic compounds, endowing it with potent antioxidant characteristics. The primary component of peppermint, "menthol," finds extensive use in the production of various products like pain balms, lozenges, cold balms, toothpaste, and more. Mint oil, along with its elements and derivatives, is employed across multiple industries, including food, flavoring, perfumery, and the pharmaceutical sector [6, 7].

1.1.1 BOTANICAL DESCRIPTION

Peppermint (*M.* × *piperita*) is an aromatic perennial herb with rhizomatous roots, typically growing to heights of 30–90 cm, and its stems are upright and four-angled. The stems, which are often branched, display a violet or purplish hue. The leaves range in length from 4 to 5 cm and are dim to light green in color. These leaves are arranged oppositely, are oblong-ovate in shape, have serrated edges, and have short petioles with serrated margins. Typically appearing in the summer, the 8 mm long, purple or scarlet flowers have a four-lobed corolla that is around 5 mm in diameter [8].

1.1.2 GEOGRAPHIC DISTRIBUTION

The peppermint species (*M.* × *piperita*) is indigenous to Europe and is abundantly present in Northern and Eastern Europe, Africa, and the United States of America. Nevertheless, it has been consistently cultivated and grown worldwide [8, 9]. Table 1.1 provides a list of the local names for peppermint in various regions across the world.

1.1.3 TAXONOMY

Peppermint is a member of the mint or sage family, *Lamiaceae*. The *Lamiaceae* family of trees and shrubs includes more than 7200 species and 260 genera. Thirteen natural hybrids and roughly 61 species make up the genus. The natural hybrid of *M. spicata* and *M. aquatica* is called *M.* × *piperita*. The species name piperita, which refers to the plant's aromatic and spicy taste, is taken from the Latin piper, which means pepper. The term "Mentha" is derived from the Greek name "Mintha," which was associated with a mythical nymph who transformed into this plant [10, 11]. The scientific classification of peppermint is illustrated in Figure 1.1.

TABLE 1.1 Local Name of Peppermint Around the World

Arabic	Nana, Bogota, Yerba Buena,
English	Brandy mint, peppermint
Chinese	Po Ho
North America	Lamb mint, Brandy mint, Lam mint, peppermint
French	Menthe, Menthe anglaise
Brazil	Nortela, pimento
Dutch	Pepermint
Spanish	Mentainglesa, Menta piperita
Danish	Pebermynte
Mexico	Menta piperita
Uruguay	Menta
Italian	Menta piperita
Russian	Myataperechnaya
Norwegian	Peppermynte
Polish	Pepparmunta
Portuguese	Hortelanapimentosa
Turkish	Nana
Hungarian	Borsus menta
Swedish	Pepparmynt
Urdu	Podina

FIGURE 1.1 Scientific classification of peppermint.

1.1.4 *TRADITIONAL USES*

Peppermint was cultivated by the Egyptians and is extensively documented in the Icelandic pharmacopeia, where it was utilized as a flavoring agent for both tea and foods [8]. In addition, dried leaves have been found in Egyptian pyramids, and it is recognized for their therapeutic benefits in traditional Chinese medicine [12]. It is frequently added to enhance the flavor and scent of the well-known "Touareg tea" in the regions of northern Africa and the Arab [8, 13]. Peppermint is used in Iranian traditional medicine for several purposes, including as a stimulant, tonic, carminative, antiviral, and antifungal. *M.* × *piperita* and its essential oils are valued for their antispasmodic, antiseptic, and aromatic qualities; they are used in traditional medicine in both Western and Eastern cultures to treat a variety of diseases, including cough, nausea, throat pain, cancer, toothaches, aches and pains, and digestive issues [14]. In the culinary and medicinal traditions of Ancient Greece, Rome, and Egypt, peppermint has a long history. Both its leaves and oil are used in traditional medicine for flavoring and cosmetic purposes. In addition, peppermint aerial parts are added to tea, which is used as a cure for conditions such as bronchitis, coughs, and throat and mouth mucosal inflammation [15].

1.2 CHEMICAL CONSTITUENTS

The main chemical elements of *M.* × *piperita* are monoterpenoids, which are also the main compounds in the plant's essential *oil*. These also include menthone (20–31%), menthol (29–48%), menthofuran (6–8%), pulegone, menthyl acetate (3–10%), pinene, piperitone, and limonene, in addition to flavonoids, caffeic acid, like luteolin and menthoside, polyphenols, like rosmaric acid, tocopherols, carotenes, eriodictyol, tannins, narirutin, choline, and betaine. Peppermint essential oil contains menthol, an important ingredient that has been highlighted in multiple studies as a key part of the plant. Menthone and menthofuran, two of the plant's metabolites, as well as pulegone, the precursor for biosynthesis, are additional components (Figure 1.2) [16, 17].

The primary biliary metabolite of menthol is menthol glucuronide, which is typically broken down in the liver and eliminated in the urine and feces. The remainder of the urine's metabolites is mono- or dihydroxylated derivatives of menthol, which are excreted as conjugates of glucuronic acid [18].

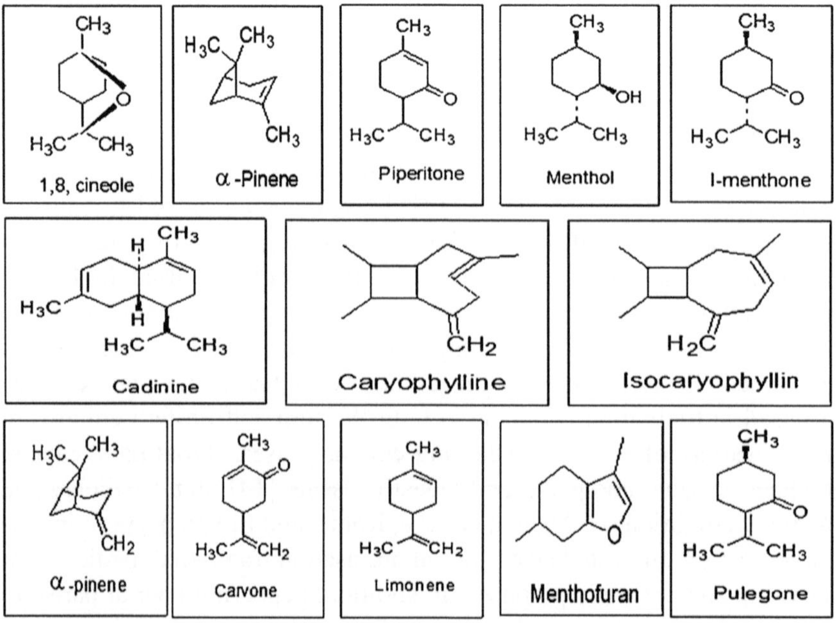

FIGURE 1.2 Various chemical constituents of peppermint oil.

In humans, 12–24 h after oral ingestion, menthol is eliminated as the glucuronide in the urine [19, 20]. Peppermint, in particular, has the ability to modulate enzymes involved in both phase I and phase II metabolism, which are crucial for the transformation and deactivation of endogenous compounds and xenobiotics. Moreover, it has demonstrated the capacity to mitigate the detrimental effects of carbon tetrachloride on liver function [21]. Despite the widespread use of peppermint as a medicinal plant in various countries, there is currently a lack of comprehensive and up-to-date scrutiny encompassing its potential toxic effects *in vitro*, as well as in animal and human studies (Figure 1.3).

1.3 PHYTOCHEMICAL CONSTITUENTS PRESENT IN PEPPERMINT

A vital class of phytoconstituents known for their well-established biological applications, including antimicrobial, anti-inflammatory, anticancer, and antioxidant, are polyphenolic components. They are classified into various subclasses, including lignans, stilbenes (2.5%), phenolic acids (42%), and flavonoids (53%).

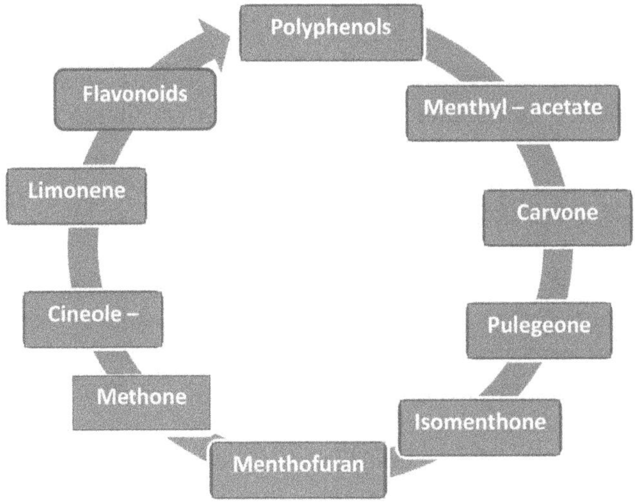

FIGURE 1.3 Major chemical components of peppermint.

1.3.1 FLAVONOIDS

Flavonoids, one of the most common polyphenols in diets, fall into six categories. *M.* × *peperita* was used to extract and identify methoxyflavones (gardenin B, 5,6-dihydroxy-7,8,3,4-tetramethoxy flavone and salvigenin), flavonols (catechin, rutin, [−]-epicatechin, quercetin, quercetin-4-glucoside, kaempferol 7-*O*-rutinoside and myricetin *O*-glucoside), and flavonones (eriodictyol, naringin, eriodictyol-glycopyranosyl rhamnopyranoside, hesperidin, and eriocitrin) [22, 23].

1.3.2 PHENOLIC ACIDS

The abundance of phenolic compounds in *M.* × *piperita* has been shown by numerous investigations. The highest quantity of caffeic acid and its metabolites out of all the phenolic acids, the latter making up 60–80% of the total phenolic compounds. It has been reported that *M.* × *piperita* contains chlorogenic acid. Rosmarinic acid is the most common phenolic in *M.* × *piperita*, making up 30% of all phenolics [24, 25].

1.3.3 LIGNANS AND STILBENES

Two lignans, medioresinol and medioresinol sulfate, and one stilbene, transresveratrol, were found in *M.* × *piperita*, according to a study [26].

1.3.4 VOLATILE COMPOUNDS

Essential oils are categorized as highly fragrant, volatile secondary metabolites of plants with a complicated molecular structure. To obtain them, aromatic and medicinal herbs are usually extracted *via* steam distillation, hydrodistillation, mechanical cold pressing, and dry distillation [27]. *M.* × *piperita* leaf has a powerful scent that is mostly made up of volatile oil. Peppermint volatile oil is a translucent yellow oil with a distinct, strong menthol scent. From peppermint, more than 300 different chemicals have been extracted and identified. Peppermint has comparatively large levels of monoterpenes (52%) and sesquiterpenoid compounds (9%). Moreover, studies have indicated that peppermint also includes lower amounts of alcohols (6%) and other substances, such as aromatic hydrocarbons (9%), aldehydes (9%), lactones (7%), and miscellaneous (8%) [28, 29].

1.4 ESSENTIAL OILS

Essential oils serve as the primary bioactive components within peppermint, known for their antiviral, antispasmodic, antimicrobial, and carminative properties. Furthermore, the essential oils extracted from various mint species are recognized as natural antioxidants. Peppermint oil is derived from either fresh garden mint leaves or dried leaves through a hydrodistillation process. It primarily consists of monoterpene and phenyl propanoids, sesquiterpene hydrocarbons, and their oxygen derivatives [30, 31]. Menthol, a major constituent of peppermint, finds wide application in addressing issues such as migraines, stuffiness, nasal congestion, and musculoskeletal pain due to its natural antiseptic and anesthetic qualities. Often referred to as “peppermint camphor,” menthol is commonly used as a soothing balm. In the realm of tobacco products, peppermint extracts are frequently added to cigarettes to mask the bitter and sour tobacco taste, providing a soothing effect on the throat without bitterness. The leaves of peppermint contain approximately 1.2–3.9% (v/w) of essential oil, housing over 300 identified compounds. The dominant class is terpenes, constituting 52% monoterpenes and 9% sesquiterpenes. Additionally, smaller percentages are attributed to aromatic hydrocarbons (9%), lactones (7%), miscellaneous compounds (8%), aldehydes (9%), and alcohols (6%) [32, 33].

The essential oils that make up peppermint range from 0.5% to 4%. Of these, around 25% to 78% are menthol, 14% to 36% are menthone, 1.5% to 10% isomenthone, 2.8 to 10% menthyl acetate, and 3.5% to 14% cineol [34]. Menthofuran, menthone, D-neoisomenthol, and pulegone were identified as the primary components in the two peppermint varieties—chocolate mint and grapefruit mint—while linalyl acetate and linalool dominated the grapefruit mint kind. The chemical composition of a commercial essential oil that was collected from India revealed that the main constituents are limonene (10.6%), isomenthone (14.8%), and menthol (19.1%) [35]. On the other hand, the composition of the peppermint oil gathered from Serbia was comparable to that of the oil gathered from Morocco, mostly consisting of menthol, menthyl acetate, and menthone. Nevertheless, substantial concentrations of menthone, isomenthone, menthyl acetate, and menthofuran were also observed, in addition to menthol (up to 59.7%). This species also has another chemotype that has been found, and it is distinguished by a high concentration of carvone, with limonene being the second principal ingredient [36, 37].

1.4.1 EXTRACTION OF POLYETHYLENE OXIDE (PEO)

For maximum essential oil extraction, mentha is harvested at the onset of flowering and grown from rhizomes. It is never easy to extract bioactive components from plants; nonetheless, there are a variety of techniques that can be used to extract essential oils from plants that have specific capabilities as active ingredients. Active compounds can be extracted from essential oils using a variety of techniques, including steam distillation, hydrogenation distillation, supercritical fluid extraction, microwave-assisted extraction, ultrasonic-assisted extraction, and countercurrent extraction [38].

The original plant raw materials and the extraction method used determine the rate of recovery and chemical makeup of essential oil components in plants, especially medicinal plants [39]. After harvesting, mentha is sun-dried and then blended into a fine powder. Following sifting, the moisture content of the plant material was evaluated gravimetrically by drying it at 105°C until a consistent weight was reached. The average particle size and particle size range were then estimated. Mass transfer limitations can be removed, and extraction will be improved by adjusting the moisture content to 8.77% and the average mentha particle size to 0.224 mm [40]. Following that, the by-products are gathered in barrels for steam distillation. The aromatic smell of essential oils can be enhanced by fractional distillation technology after hydrodistillation and ethanol solid–liquid extraction are used to extract the essential oil components [41, 42].

Menthol, menthone, neomenthol, and isomenthone are the principal chemical constituents of mentha. One of three techniques is used to extract the constituents of the essential oil: steam distillation, microwave distillation, or hydrodistillation [43]. The usual Soxhlet and supercritical fluid extraction, microwave-assisted extraction, and ultrasonic-assisted extraction are the most conventional techniques for obtaining essential oils from mentha. Using varying power levels can boost PEO yield. Conventional hydrodistillation is a historical approach that has been demonstrated to separate the components of mentha essential oil, making it a viable alternative method. Hydrodistillation with microwave assistance is another method for separating PEO [44]. Microwave-assisted distillation can cut the extraction time from 180 to 30 min, even though steam distillation and hydrodistillation offer higher extraction yields and quantitation of important components.

1.5 APPLICATIONS OF PEPPERMINT OIL

Traditional uses of piperita include colds, fever, digestive disorders, oral mucosa irritation, and antifungal and antiviral properties. The biological qualities of pepper oil, such as its antimicrobial, antioxidant, biopesticidal, antiviral, anticancer, anti-inflammatory, larvicidal, radioprotective effect, genotoxicity, and antidiabetic activity, have come to light as a result of scientific research. Mentha piperita, also referred to as peppermint, has been used extensively in both Eastern and Western nations as a traditional medicine for irritation, inflammation, and the production of toothpaste and mouthwashes [45]. It is also beneficial for respiratory disorders as an anticongestive and expectorant, as well as for the digestive and vascular systems as an antispasmodic. Traditional uses include larvicidal and insect repellent properties, as well as irritable bowel syndrome in humans. Egyptians grew *M. piperita*, which was well documented in the Icelandic pharmacopoeia as a flavoring agent in dishes and tea. *M. piperita* has been used therapeutically in Chinese traditional medicine, and its dried leaves have been discovered in Egyptian pyramids [46]. In the Arab world and northern Africa, peppermint is frequently added to popular drinks like "Touareg tea" as a flavoring and scent. Peppermint is used as a carminative, stimulant, tonic, antiviral, and antifungal in traditional medicine. *M. piperita* and its essential oils are used in both Western and Eastern traditional medicine as an aromatic, anti-inflammatory, antispasmodic, and antiseptic agent. They are also used to treat tumors, toothaches, cramps, indigestion, colds, nausea, and sore throats. Ancient Greek, Roman, and Egyptian societies used peppermint for both culinary and medicinal purposes [47]. Traditional medicine uses peppermint leaves and oil as a flavoring and cosmetic. Tea infused with peppermint oil is a good remedy for bronchitis, coughs, and irritation of the throat and oral mucosa. Traditional medicine uses the leaf's essential oil to treat rheumatism, muscle soreness, and menstrual cramps. This *M. piperita* leaf decoction is used as a home treatment for inflammation, bronchitis, and coughs. *M. piperita* is widely used in oral health care and cosmetic products due to its potent aromatic smell and numerous biological and physiological benefits [48]. The British Herbal Compendium lists peppermint leaves as an official ingredient in a remedy for intestinal colic, dyspepsia, biliary disorders, and flatulence. Peppermint is suggested by the European Scientific Cooperative on Phytotherapy as a remedy for intestinal issues and enteritis. The German Standard Licence also makes reference to peppermint leaf teas, which are used to alleviate gastrointestinal issues and gallbladder issues [49]. Traditional medicine has utilized the medicinal preparation of this well-known herbal tea (33% chamomile flowers and 67% peppermint leaf) to treat stomach ulcers and gastrointestinal discomfort. For the treatment of coryza and nasal catarrh in children, inhalation (10% peppermint oil + 45% eucalyptus oil + 45% pumilio pine oil) is recommended. It has several uses in traditional American medicine, such as a carminative in antacids, an antipruritic in sunburn creams, a counterirritant in topical analgesics, a decongestant in inhalants and lozenges, and an antiseptic or flavoring ingredient in mouthwash, gums, and toothpaste [50].

1.6 BIOLOGICAL ACTIVITIES OF PEO

PEO is primarily metabolized within the body through its interaction with glucuronic acid, subsequently being excreted in either urine or feces. A substantial body of research underscores the multiple beneficial attributes of PEO, including anti-inflammatory and antiallergenic properties, antibacterial and antioxidant effects, as well as antifungal, antiviral, antiradiation, antitumor, and neuroprotective activities, as illustrated in Figure 1.4.

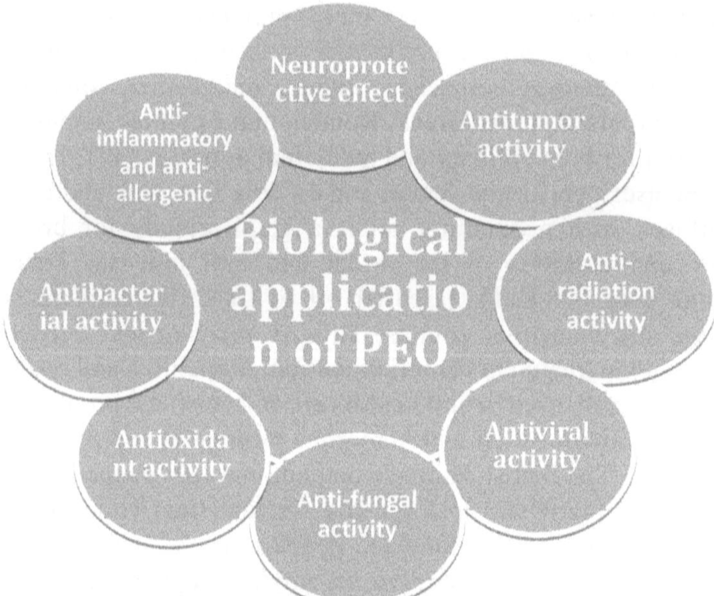

FIGURE 1.4 Biological application of peppermint essential oil.

1.6.1 *ANTI-INFLAMMATORY AND ANTIALLERGENIC ACTIVITY*

It has been observed that peppermint oil containing menthol and cineole reduces homologous cutaneous anaphylaxis in guinea pigs, a condition brought on by IgE antibodies. This study indicates that peppermint has a significant anti-inflammatory effect. In *in vivo* studies, Kamei et al. [51] and Inoue et al. [52] administered ethanol–peppermint (*M. piperita*) extract to rats suffering from nasal allergies at specific nasal dosages of 300 mg and 1000 mg, respectively, and saw a reduction in sneezing [51, 52]. Additionally, Inoue et al. [53] investigated the flavonoids, primarily peppermint's luteolin-7-*O*-rutinoside, and discovered that it had an inhibitory effect on allergies mediated by antigens in the nasal cavity [53].

1.6.2 ANTIBACTERIAL ACTIVITY

Iscan et al.'s (2002) study examined the effects of peppermint oil and menthol, a component of it, on nearly 21 different types of pathogens. They discovered that menthol had inhibitory effects on both human and animal pathogens, such as *Staphylococcus aureus* and *Listeria monocytogenes*, as well as on some plant pathogens like *Xanthomonas* and *Pseudomonas* strains [54]. According to Tassou et al. (2000), peppermint oil has a protective effect against *Enterococcus faecium* and *S. aureus* as well as a favorable effect on lowering the count of colony-forming units of these bacteria [55].

1.6.3 ANTIOXIDANT ACTIVITY

Mekinic et al. (2014) investigated the ethanol/water extract of peppermint leaves at a specific ratio of 80:20 and found that peppermint extract is the most effective in preventing unintentional *β*-carotene bleaching [56]. According to Li et al. (2017), peppermint ethanol extract strongly inhibited the DPPH (2,2-diphenyl-1-picrylhydrazyl) radical in their experiment [57]. After conducting additional research, Farnad et al. (2014) confirmed the results of their experiment and found that the ethanol extract of peppermint leaves (*M. piperita*) reduced the capacity of nitric oxide and DPPH to scavenge radicals [58]. Mallick et al. (2016) looked into the menthol of peppermint extract's *in vitro* antioxidant activity and discovered a sizable scavenging activity on DPPH radicals [59]. A study by Mimica-Dukic et al. [60] found that peppermint (*M. piperita*) has the strongest potential to block DPPH radical by reducing the generation of the OH radical.

1.6.4 ANTIFUNGAL ACTIVITY

Nilo et al. (2017) reported that the peppermint leaf extract exhibits fungicidal properties that prevent the growth of specific fungi, such as *Aspergillus niger*, *Fusarium moniliforme*, and *Aspergillus fumigatus* [23].

Peppermint oil has been shown to have an inhibiting impact on *A. niger*, *A. flavus*, *Rhizopus solani*, and *Fusarium solani*, according to Hussain et al. [61]. Similarly, according to Sandasi et al. (2011), peppermint oil was discovered to prevent *Candida albicans* from growing [62]. A plethora of noteworthy literature exists regarding peppermint and its ability to act as a fungicidal agent against several species of fungi, with established mechanisms of action. Many antifungal therapies for humans have been proposed using peppermint [63–65].

1.6.5 ANTIVIRAL ACTIVITY

The peppermint leaf aqueous extract is highly effective at inhibiting several viruses, such as influenza A, herpes simplex virus (HSV), vaccine virus, and many others that are mostly found in eggs and the cell-culture system. It was discovered that peppermint aqueous extract may be effective against HIV-1 (human immunodeficiency virus) [66]. According to Minami et al. (2003), peppermint oil (1%) significantly inhibits the growth of HSV-1-infected cells and has a strong effect against HSV-1 [67]. Peppermint extract, which primarily consists of dichloromethane and methanol, has been shown in another study by Yucharoen et al. (2012) to have a strong inhibitory impact against both HSV-1 and HSV-2 [68]. According to a different study conducted in 2006 by Nolkemper et al., peppermint had the strongest (85%) protective effect against HSV-1 and HSV-2 [69].

1.6.6 ANTIRADIATION ACTIVITY

It was discovered that the peppermint (*M. piperita*) extract exhibited a preventive effect against chromosomal damage caused by radiation. When mice were given an aqueous peppermint extract dosage of 1 g/kg body weight orally. Samarth et al. (2001) examined an *in vivo* experiment using mice, they found that the animals' leukocyte count, erythrocyte count, hemoglobin level, and spleen colonies all rose [70]. In an *in vivo* study using mice, Samarth et al. (2004) found that the antiradiation characteristics of peppermint essential oil caused a change in the hematological state of the blood. Merely 17% of the mice in the test group that received peppermint oil (menthol oil) died due to radiation damage, whereas 100% of the mice in the control group perished due to radiological effects [71]. Furthermore, Samarth and Samarth (2009) showed a substantial reduction in lipid peroxidation in mice given oral doses of *M. piperita* leaf extract for 30 days [72].

1.6.7 ANTITUMOR ACTIVITY

According to one study (Ohara and Matsuhisa, 2002), about 120 medicinal herbs have antitumor action. It was shown that these inhibited the tumor promoter genes linked to tumorigenic activity, including okadaic acid [73]. Among these is peppermint, which is the most commonly eaten edible herb that, as previously mentioned [74], has an antitumor impact by reducing okadaic acid. According to another study, menthol, the major constituent of peppermint, impacts the enzyme of *N*-terminal acetyltransferases, which is principally involved in xenobiotic metabolism by transferring the acetyl group from acetyl-CoA to its xenobiotic acceptor substrates [75]. Thus, it preferentially reduces tumor formation by inactivating tumor lining cells in the liver [76].

1.6.8 NEUROPROTECTIVE EFFECT

Phasic and tonic GABAA (gamma-aminobutyric acid A) receptor-mediated neuronal currents across the periaqueductal grey (PAG) are enhanced by menthol in PEO. It seems that menthol influences tonic current through extrasynaptic GABAA receptors that lack the delta subunit. The analgesic and anxiolytic effects of this brain region can be altered by menthol-induced increases in GABAergic inhibition within the PAG [77]. When it comes to the kinetics of IPSC in hippocampal CA1 pyramidal neurons, menthol has no effect, but it slows the decay of IPSC in PAG neurons. Moreover, menthol significantly inhibits acetylcholinesterase and binds to GABAA and nicotinic receptors in a concentration-dependent manner, which may increase synaptic availability of acetylcholine [78]. PEO protects CAD cells from oxidative stress by increasing Ca^{2+} concentration and prolonging the depolarization response. Additionally, menthol has a significant impact on subpopulations of sensory ganglion neurons. Menthol enhances the perception of cooling and enhances the ability of isolated sensory neurons to respond to temperature reduction. The results show that whereas menthol has no effect on oral trigeminal temperature sensors, it can suppress trigeminal neurons that sense cold [79].

1.7 PHARMACOLOGICAL EFFECTS OF PEO

It is widely believed that PEO has a variety of pharmacological effects, including antihepatotoxicity, nausea prevention, neuropsychiatric effects, diabetes, wound healing, and psoriasis, liver, kidney, and respiratory protection, fever reduction, COVID-19, and peppermint. Understanding these pharmacological effects mechanistically is beneficial to PEO use (Figure 1.5).

FIGURE 1.5 Pharmacological application of peppermint essential oil.

1.7.1 ANTIHEPATOTOXICITY EFFECTS

The majority of chemical reactions in the liver result in the production of numerous toxic substances, including carbon tetrachloride (CCl_4), uric acid, creatinine, oxygen radicals, and other substances that damage the liver. Some viruses, medications, and other substances can also be major contributors to liver damage, also known as hepatotoxicity. According to a study by Khalil et al. (2015), peppermint leaves are effective at protecting against a variety of hazardous substances produced in the liver, including CCl_4. The study was conducted *in vivo* on rat models with an oral peppermint oil dose of 0.5 ml/kg. CCl_4 was assumed to cause hepatotoxicity by lowering the levels of alanine transferase, aspartate aminotransferase, alkaline phosphatase, bilirubin, urea, creatinine, uric acid, and specific peroxidases, such as glutathione peroxidase, all of which induce hepatotoxicity [80]. Maliakal and Wanwimolruk (2001) claim that aqueous peppermint extract has the ability to modulate several phases of drug metabolism enzymes. It contains phenolic compounds and carotenoids, which have the potential to absorb free radicals [81]. According to other studies, plant extract's anthocyanin can shield the liver from lipoperoxidation brought on by carbon tetrachloride [21]. Peppermint leaves have been shown in another study by Ogaly et al. (2018) to have a high antihepatotoxicity effect through enhancing lipid peroxidation, transforming growth factor-1, and decreasing desmin [82].

1.7.2 NAUSEA PREVENTING ACTIVITIES

Peppermint tea's antispasmodic qualities relieve nausea and prevent puking. The smell of peppermint can exacerbate nausea, but the plant's chemical components help to relax the muscles that compress the stomach, preventing retching. If one is inebriated before boarding an airplane, boat, or ship, it is similarly beneficial to cure motion sickness or sea sickness [80].

1.7.3 NEUROPSYCHIATRIC EFFECTS

Some scientists claim that peppermint is well known for stimulating the central nervous system. Based on possible alterations in brain activity, they studied how fragrances affected pain perception, perceived physical exertion, and cognitive performance [83].

1.7.4 PEPPERMINT IN DIABETES

According to Hemalakshmi et al. (2012), diabetes is a metabolic disorder characterized by elevated serum glucose levels brought on by either insulin inactivation or a lack of the glucose-anabolic enzyme, insulin. The disorder also manifests as abnormalities in the metabolism of fats, proteins, and carbohydrates, which upset homeostasis and cause a host of physiological issues [84]. *In vivo* effects of a 2 mM peppermint-salicylic acid infusion extract on blood and urine glucose, albumin, urea, and uric acid levels in diabetic rats were studied by Figueroa-Pérez et al. (2018) [85]. Barbalho et al. (2011) reported that the extract significantly reduced blood glucose levels while simultaneously increasing insulin levels in the blood. In this work, diabetogenic rats that had received an injection of the glucosamine-nitrosourea combination streptozotocin, as well as non-diabetogenic rats, were administered an oral dose of peppermint leaf extract (0.29 g/kg; 100 g/L) [86]. Nevertheless, the complete mechanism by which peppermint lowered blood glucose levels was not fully known. Furthermore, Badal et al. (2011) demonstrated through their fructose-fed male Sprague Dawley rats that the administration of 100 and 250 mg/kg of peppermint oil per day can lower blood levels of triglycerides, cholesterol, very low-density lipoprotein, low-density lipoprotein (LDL), and serum glucose [87].

1.7.5 TRANSDERMAL MEDICATION FOR WOUND HEALING AND PSORIASIS

The process of healing wounds is intricate and involves various processes such as fibroblast proliferation, neovascularization, angiogenesis, and epidermal regeneration [88]. The innate immune system is frequently activated both locally and systemically by severe burn wounds, which can cause significant inflammation, immunosuppression, and damage to multiple organs. For example, burn wounds have the potential to enhance IL-10 levels and suppress the Th1 response, both of which are linked to an increased risk of infection [89]. A proponent of dermal fibrosis, transforming growth factor beta (TGF-β) has a variety of functions in the burn wound. In the later stages of the healing process, TGF-β may act as a factor that promotes wound healing and amplifies overhealing consequences, such as hypertrophic scar and keloid [90].

1.7.6 FEVER REDUCING

Fever relief is one of peppermint's unintended benefits, particularly when infused with tea. The natural cooling impact of menthol on the body could be the cause of this. The body goes through detoxification in order to rule out other causes of fever, cough, cold, and flu. The menthol reduces fever by acting internally at the same time. Furthermore, menthol naturally soothes muscles, which helps lessen the aches and pains brought on by a high body temperature [91].

1.7.7 LIVER, KIDNEY, AND RESPIRATORY PROTECTION

PEO reduces intrahepatic cholestasis and is choleretic. By possibly controlling the expression of certain genes involved in bile acid synthesis, PEO can raise bile acid concentration, decrease total cholesterol, and encourage bile secretion. In PEO, menthol and menthone in particular can influence inflammatory processes, decrease liver fibrosis, diminish the coupling of cholesterol secretion to bile salt secretion, and enhance bile flow [92]. The development of stones in the urinary system is known as urolithiasis. Urinary crystals and elevated calcium oxalate concentrations cause inflammation of the tubular epithelium, which is one of the main risk factors for urolithiasis.

Menthol suppresses respiratory sensory irritation response to various cigarette smoke irritants because it can abolish the mouse irritation response to acrolein, an agonist of the transient receptor potential ankyrin 1. Menthol is an agonist of the TRPM8 ion channel in cold-sensitive sensory neurons. Furthermore, the presence of menthol can lessen the discomfort caused by cyclohexanone, acetic acid, and transient receptor potential cation channel subfamily V member 1. These findings indicate that menthol is well absorbed in the respiratory system and that it generates a local concentration that is sufficient to activate sensory TRP channels [93].

1.7.8 COVID-19 AND PEPPERMINT

Menthone and menthol, two bioactive chemicals found in peppermint essential oils, have the ability to prevent the growth of certain viruses, including the COVID-19 virus. They can also assist in enhancing a person's respiratory system by opening up their bronchioles and facilitating the breakdown of mucus [94]. The phytochemicals in peppermint (*M. piperita* L.) have been shown to suppress the capacity of viruses such as SARS-CoV to spread through a variety of methods. A few of these methods included breaking through the viral membrane, directly inhibiting the virus, stopping its replication, and deactivating its several enzymes [95]. Primarily through the formation of the complex "ACE2 receptor (of the host body)–S1 (region of the spike protein of the COVID-19 virus)." A *silico* study by Kulkarni et al. (2020) discovered that the phytochemicals of essential oils of various medicinal herbs, such as peppermint's "pulegeone" and "menthone," inhibit the actionability of the COVID virus's spike protein by deactivating the region of the protein known as S-1, which is normally occupied by the host body's ACE-2 receptor. As a result, peppermint stops the COVID-19 virus from interacting with the host [96]. COVID-19 virus infections significantly affect the respiratory system. Researchers have discovered that peppermint menthol, through the formation of "Cold-Menthol-Receptor-complexes," is extremely effective in avoiding chronic obstructive pulmonary disease. In addition to peppermint vapor, hot peppermint tea and other products containing peppermint oil provide assistance from a variety of nasal issues as previously mentioned. Studies have demonstrated that peppermint is beneficial in post-COVID situations as well. Given that peppermint has the ability to enhance flavor, this enhances taste perception, which is a significant difficulty during this viral infection [97].

1.8 CONCLUSION

PEO is a blend of mentha's volatile secondary metabolites; it has a cool, slightly bitter flavor and a distinct mint scent. PEO is widely recognized to consist primarily of menthol, menthone, neomenthol, and isomenthone. These constituents exhibit potent anti-inflammatory and antiallergenic properties, as well as antibacterial, antioxidant, antifungal, antiviral, antiradiation, antitumor, and neuroprotective effects. Additionally, PEO is known to have antipsychiatric effects, and to be effective against diabetes, wound healing, psoriasis, liver, kidney, and respiratory protection, fever-reducing, COVID-19, and peppermint.

Mentha extraction and PEO isolation are challenging but essential steps in the downstream investigation of PEO, which includes pharmacological impact determination, biological activity assessment, and application evaluation. Four distinct types of components have currently been identified in

PEO: menthol, menthone, neomenthol, and isomenthone. Nevertheless, in order to support the downstream analysis of PEO, some other novel bioactive chemicals in PEO should be found employing cutting-edge methods, including supercritical fluid extraction, membrane separation technology, and semi-bionic extraction methodology.

There is a pressing need for a thorough investigation using cutting-edge techniques to unravel the underlying mechanisms. Specifically, it is crucial to gain a deeper understanding of the molecular targets and the signaling pathways affected by PEO. This exploration may uncover novel targets and the subsequent sequences influenced by PEO. Notably, the transdermal absorption observed after topical PEO administration is distinctive, enabling convenient treatment for both local conditions and systemic disorders once PEO enters the bloodstream. While mentha and PEO generally offer high therapeutic benefits with few side effects, concerns about PEO toxicity at higher doses remain a significant public issue. It is essential to ascertain the toxic dosage under specific circumstances and understand the toxic mechanisms to ensure the safe application of PEO across various domains.

KEYWORDS

- **peppermint**
- **essential oil**
- **chemical constituents**
- **quality characteristics**

REFERENCES

1. Dhakad, A.K., Ikram, M., Sharma, S., Khan, S., Pandey, V.V., Singh, A. A biological, nutritional, and therapeutic significance of Moringa oleifera Lam. *Phytotherapy Research*, 2019, *33*, 2870–2903.
2. Loolaie, M., Moasefi, N., Rasouli, H., Adibi, H. Peppermint and its functionality: A review. *Archives of Clinical Microbiology*, 2017, *8*(4), 54.
3. Shetta, A., Kegere, J., Mamdouh, W. Comparative study of encapsulated peppermint and green tea essential oils in chitosan nanoparticles: Encapsulation, thermal stability,

in-vitro release, antioxidant and antibacterial activities. *International Journal of Biological Macromolecules*, 2019, *126*, 731–742.

4. Singh, R., Shushni, M.A.M., Belkheir, A. Antibacterial and antioxidant activities of Mentha piperita L. *Arabian Journal of Chemistry*, 2015, *8*(3), 322–328.
5. Loolaie, M., Moasefi, N., Rasouli, H., Adibi, H. Peppermint and its functionality: A review. *Archives of Clinical Microbiology*, 2017, *8*(4), 54.
6. Mahendran, G., Rahman, L. Ethnomedicinal, phytochemical and pharmacological updates on Peppermint (Mentha × piperita L.)—A review. *Phytotherapy Research: PTR*, 2020, *34*(9), 2088–2139.
7. Ranjbar, M., Kiani, M., Nikpay, A. Antioxidant and scolicidal activities of four Iranian Mentha species (Lamiaceae) in relation to phenolic elements. *Journal of Herbmed Pharmacology*, 2020, *9*(3), 200–208.
8. Singh, R., Shushni, A.M.M., Belkheir, A. Antibacterial and antioxidant activities of Mentha piperita L. *Arabian Journal of Chemistry*, 2015, *8*, 322–328.
9. Singh, P., Pandey, A.K. Prospective of essential oils of the genus Mentha as biopesticides: A review. *Frontiers in Plant Science*, 2018, *9*, 1295.
10. Benabdallah, A., Boumendjel, M., Aissi, O., Rahmoune, C., Boussaid, M., Messaoud, C. Chemical composition, antioxidant activity and acetyl cholinesterase inhibitory of wild Mentha species from Northeastern Algeria. *South African Journal of Botany*, 2018, *116*, 131–139.
11. Kumar, P., Mishra, S., Malik, A., Satya, S. Insecticidal properties of Mentha species: A review. *Industrial Crops and Products*, 2011, *34*, 802–817.
12. Beigi, M., Torki-Harchegani, M., Pirbalouti, A.G. Quantity and chemical composition of essential oil of peppermint (Mentha×piperita L.) leaves under different drying methods. *International Journal of Food Properties*, 2018, *21*(1), 267–276.
13. Singh, P., Pandey, A.K., Tripathi, N.N. Essential oils: A renewable source for the management of stored product insects—A review. *Agricultural Reviews*, 2012, *33*, 226–236.
14. Herro, E., Jacob, S.E. Mentha piperita (peppermint). *Dermatitis*, 2010, *21*(6), 327–329.
15. Balakrishnan, A. Therapeutic uses of Peppermint—A review. *Journal of Pharmaceutical Sciences and Research*, 2015, *7*(7), 474–476.
16. Orio, L., et al. Hydrodistillation and in situ microwave-generated hydrodistillation of fresh and dried mint leaves: A comparison study. *Journal of the Science of Food and Agriculture*, 2012, *92*(15), 3085–3090.
17. Radivojac, A., et al. Extraction of peppermint essential oils and lipophilic compounds: assessment of process kinetics and environmental impacts with multiple techniques. *Molecules (Basel, Switzerland)*, 2021, *26*(10), 2879.
18. Uribe, E., Marín, D., Veg Gálvez, A., Quispe-Fuentes, I., Rodríguez, A. Assessment of vacuum-dried peppermint (mentha piperita l.) as a source of natural antioxidants. *Food Chemistry*, 2016, *190*, 559–565.
19. Riachi, L.G., De Maria, C.A.B. Peppermint antioxidants revisited. *Food Chemisty*, 2015, *176*, 72–81.
20. Peat, J., Frazee, C., Kearns, G., Garg, U. Determination of menthol in plasma and urine by gas chromatography/mass spectrometry (gc/ ms). *Methods in Molecular Biology*, 2016, *1383*, 205–211.

21. Arab Ameri, S., Samadi, F., Dastar, B., Zerehdaran, S. Effect of peppermint (mentha piperita) powder on immune response of broiler chickens in heat stress. *Iranian Journal of Applied Animal Sciences*, 2016, *6*, 435–445.
22. Pandey, K.B., Rizvi, S.I. Plant polyphenols as dietary antioxidants in human health and disease. *Oxidative Medicine and Cellular Longevity*, 2009, *2*, 270–278.
23. Nilo, M.C.S., Riachi, L.G., Simas, D.L.R., Coelho, G.C., da Silva, A.J.R., Costa, D.C.M., Alviano, D.S., Alviano, C.S., De Maria, C.A.B. Chemical composition and antioxidant and antifungal properties of Mentha × piperita L. (Peppermint) and Mentha arvensis L. (cornmint) samples. *Food Research*, 2017, *1*(5), 147–156.
24. Fecka, I., Turek, S. Determination of water-soluble polyphenolic compounds in commercial herbal teas from Lamiaceae: Peppermint, Melissa and Sage. *Journal of Agricultural and Food Chemistry*, 2007, *55*(26), 10908–10917.
25. Misan, A.C., Mimica-Dukic, N., Mandic, A.I., Sakac, M.B., Milovanovic, I.L., Sedej, I.J. Development of a rapid resolution HPLC method for the separation and determination of 17 phenolic compounds in crude plant extracts. *Central European Journal of Chemistry*, 2011, *9*(1), 133–142.
26. Mekinic, I.G., Skroza, D., Ljubenkov, I., Simat, V., Mozina, S.S., Katalinic, V. In vitro antioxidant and antibacterial activity of Lamiaceae phenolic extracts: A correlation study. *Food Technology and Biotechnology*, 2014, *52*(1), 119–127.
27. Figueiredo, A.C., Barroso, J.G., Pedro, L.G., Scheffer, J.C. Factors affecting secondary metabolite production in plants: Volatile components and essential oils. *Flavour and Fragrance Journal*, 2008, *23*, 213–226.
28. Brahmi, F., Khodir, M., Mohamed, C., Pierre, D. Chemical composition and biological activities of Mentha species. InTech: In: *Aromatic and Medicinal Plants-Back to Nature*. 2017.
29. Riachi, L.G., De Maria, C.A. Peppermint antioxidants revisited. *Food Chemistry*, 2015, *176*, 72–81.
30. Solórzano-Santos, F., Miranda-Novales, M.G. Essential oils from aromatic herbs as antimicrobial agents. *Current Opinion in Biotechnology*, 2012, *23*, 136–141.
31. Aljaafari, M., Alhosani, M.S., Abushelaibi, A., Lai, K.-S., Lim, S.-H.E. *Essential Oils: Partnering with Antibiotics. In Essential—Oils of Nature*. IntechOpen: London, UK, 2019.
32. Srivastava, A.K., Singh, V.K. Biological action of essential oils (Terpenes). *Biomedscidirect*, 2019, *10*, 6854–6859.
33. Gupta, A.K., Rather, M.A., Kumar Jha, A., Shashank, A., Singhal, S., Sharma, M., Pathak, U., Sharma, D., Mastinu, A. *Artocarpus lakoocha* Roxb. and *Artocarpus heterophyllus* Lam. Flowers: New sources of bioactive compounds. *Plants*, 2020, *9*, 1329.
34. Patil, S.R., Patil, R.S., Godghate, A.G. Mentha piperita Linn: Phytochemical, antibacterial and diterian adulticidal approach. *International Journal of Pharmacy and Pharmaceutical Sciences*, 2016, *8*(3), 352–355.
35. Pulipati, S., SaiKoushik, O., Babu, P.S. Phytochemical analysis and antibacterial efficacy of Mentha piperita (L) ethanolic leaf extract against clinical isolates of uropathogens. *British Microbiology Research Journal*, 2016, *13*(6), 1–5.

36. Rahdari, T., Hamzei, M. Repellency effect of essential oils of Mentha piperita, Rosmarinus officinalis and Coriandrum sativum on *Tribolium confusum* Duval (Coleoptera: Tenebrionidae). *Journal of Chemical Research*, 2017, *2*(2), 107–112.
37. Saeidi, K., Mirfakhraie, S. Chemical composition and insecticidal activity Mentha piperita L. essential oil against the cowpea seed beetle Callosobruchus maculatus F. (Coleoptera: Bruchidae). *Journal of Entomological and Acarological Research*, 2017, *49*, 67–69.
38. Messaoudi, M., et al. Mentha pulegium effect of extraction methods on polyphenols, flavonoids, mineral elements, and biological activities of essential oil and extracts of L, *Molecules (Basel, Switzerland)*, 2021, *27*(1), 11.
39. Spadaccino, G., et al., Essential oil characterization of Prunus spinosa L., Salvia officinalis L., Eucalyptus globulus L., Melissa officinalis L. and Mentha × piperita L. by a volatolomic approach. *Journal of Pharmaceutical and Biomedical Analysis*, 2021, 202, 114167.
40. Pavlić, B., et al. Antioxidant and enzyme-inhibitory activity of peppermint extracts and essential oils obtained by conventional and emerging extraction techniques. *Food Chemistry*, 2021, 338, 127724.
41. Cohen, S., et al. FEMA GRAS assessment of natural flavor complexes: Mint, buchu, dill and caraway derived flavoring ingredients. *Food and Chemical Toxicology*, 2020, 135, 110870.
42. Orio, L., et al. Hydrodistillation and in situ microwave-generated hydrodistillation of fresh and dried mint leaves: A comparison study, *Journal of the Science of Food and Agriculture,* 2012, 92(15), 3085–3090.
43. Radivojac, A., et al. Extraction of peppermint essential oils and lipophilic compounds: Assessment of process kinetics and environmental impacts with multiple techniques. *Molecules (Basel, Switzerland)*, 2021, *26*(10), 2879.
44. Park, H., et al. Simple preparative gas chromatographic method for isolation of menthol and menthone from peppermint oil, with quantitative GC-MS and (1) H NMR assay. *Journal of Separation Science*, 2012, 35(3), 416–423.
45. Kamatou, G., et al. Menthol: A simple monoterpene with remarkable biological properties. *Phytochemistry*, 2013, *96*, 15–25.
46. Boon, H., Smith, M., *The Complete Natural Medicine Guide to the 50 Most Common Medicinal Herbs*. Robert Rose, 2004.
47. Mahendran, G., Rahman, L.U. Ethnomedicinal, phytochemical and pharmacological updates on Peppermint (Mentha× piperita L.)—A review. *Phytotherapy Research,* 2020, *34*(9), 2088–2139.
48. Kaur, C., Saraf, S. In vitro sun protection factor determination of herbal oils used in cosmetics. *Pharmacognosy Research*, 2010, 2(1), 22–25.
49. Elsaie, L., et al. Effectiveness of topical peppermint oil on symptomatic treatment of chronic pruritus. *Clinical, Cosmetic and Investigational Dermatology*, 2016, *9*, 333–338.
50. Mani Badal, R., et al. Pharmacological action of mentha piperita on lipid profile in fructose-fed rats. *Iranian Journal of Pharmaceutical Research*, 2011, *10*(4), 843–848.
51. Kamei, C., Inoue, T., Sugimoto, Y., Masuda, H. Effects of peppermint extracts on experimental allergic rhinitis in rats. *Aroma Research,* 2000, *1*(3), 61–66.

52. Inoue, T., Sugimoto, Y., Masuda, H., Kamei, C. Effects of peppermint (Mentha piperita L.) extracts on experimental allergic rhinitis in rats. *Biological and Pharmaceutical Bulletin,* 2001, *24*(1), 92–95.
53. Inoue, T., Sugimoto, Y., Masuda, H., Kamei, C. Antiallergic effect of flavonoid glycosides obtained from Mentha piperita L. *Biological and Pharmaceutical Bulletin,* 2002, *25*(2), 256–259.
54. İşcan, G., Kirimer, N., Kürkcüoğlu, M.N., Başer, H.C., Demirci, F. "Antimicrobial screening of Mentha piperita essential oils. *Journal of Agricultural and Food Chemistry*, 2002, *50*(14), 3943–3946.
55. Tassou, C., Koutsoumanis, K., Nychas, G.-J. Inhibition of Salmonella enteritidis and Staphylococcus aureus in nutrient broth by mint essential oil. *Food Research International,* 2000, *33*(3–4), 273–280.
56. Generalić Mekinić, I., Skroza, D., Ljubenkov, I., Šimat, V., Smole Možina, S., Katalinić, V. In vitro antioxidant and antibacterial activity of Lamiaceae phenolic extracts: A correlation study. *Food Technology and Biotechnology,* 2014, *52*(1), 119–127.
57. Li, Y., Liu, Y., Ma, A., Bao, Y., Wang, M., Sun, Z. In vitro antiviral, anti-inflammatory, and antioxidant activities of the ethanol extract of Mentha piperita L. *Food Science and Biotechnology,* 2017, *26*, 1675–1683.
58. Farnad, N., Heidari, R., Aslanipour, B. Phenolic composition and comparison of antioxidant activity of alcoholic extracts of Peppermint (Mentha piperita). *Journal of Food Measurement and Characterization,* 2014, *8*, 113–121.
59. Mallick, B., Sinha, S., Roy, D. Evaluation of antioxidative potential of field grown and tissue culture derived Mentha piperita L. plants. *International Journal of Current Microbiology and Applied Sciences,* 2016, 5(3), 382–391.
60. Mimica-Dukić, N., Božin, B., Soković, M., Mihajlović, B., Matavulj, M. Antimicrobial and antioxidant activities of three Mentha species essential oils. *Planta Medica,* 2003, *69*(5), 413–419.
61. Hussain, A.I., Anwar, F., Nigam, P.S., Ashraf, M., Gilani, A.H. Seasonal variation in content, chemical composition and antimicrobial and cytotoxic activities of essential oils from four Mentha species. *Journal of the Science of Food and Agriculture,* 2010, *90*(11), 1827–1836.
62. Sandasi, M., Leonard, C., Van Vuuren, S., Viljoen, A. Peppermint (Mentha piperita) inhibits microbial biofilms in vitro. *South African Journal of Botany,* 2011, *77*(1), 80–85.
63. Hamza, O.J., et al. Antifungal activity of some Tanzanian plants used traditionally for the treatment of fungal infections. *Journal of Ethnopharmacology,* 2006, *108*(1), 124–132.
64. Blaszczyk, T., Krzyzanowska, J., Lamer-Zarawska, E. Screening for antimycotic properties of 56 traditional Chinese drugs. *Phytotherapy Research,* 2000, *14*(3), 210–212.
65. Ezzat, S. In vitro inhibition of Candida albicans growth by plant extracts and essential oils. *World Journal of Microbiology and Biotechnology,* 2001, *17*, 757–759.
66. McKay, D.L., Blumberg, J.B. A review of the bioactivity and potential health benefits of peppermint tea (Mentha piperita L.). *Phytotherapy Research: An International*

Journal Devoted to Pharmacological and Toxicological Evaluation of Natural Product Derivatives, 2006, *20*(8), 619–633.

67. Minami, M., Kita, M., Nakaya, T., Yamamoto, T., Kuriyama, H., Imanishi, J. The inhibitory effect of essential oils on herpes simplex virus type-1 replication in vitro. *Microbiology and Immunology,* 2003, *47*(9), 681–684.
68. Yucharoen, R., Meepowpan, P., Tragoolpua, Y. Inhibitory effect of peppermint extracts and menthol against Herpes Simplex virus infection. *Chiang Mai Journal of Science,* 2012, *39*(1), 97–110.
69. Nolkemper, S., Reichling, J., Stintzing, F.C., Carle, R., Schnitzler, P. Antiviral effect of aqueous extracts from species of the Lamiaceae family against Herpes simplex virus type 1 and type 2 in vitro. *Planta Medica,* 2006, *72*(15), 1378–1382.
70. Samarth, R., Goyal, P., Kumar, A. Modulatory effect of Mentha piperita (Linn.) on serum phosphatases activity in Swiss albino mice against gamma irradiation. *Indian Journal of Experimental Biology*, 2001, *39*(5), 479–482.
71. Samarth, R., Goyal, P., Kumar, A. Protection of swiss albino mice against whole-body gamma irradiation by Mentha piperita (Linn.). *Phytotherapy Research: An International Journal Devoted to Pharmacological and Toxicological Evaluation of Natural Product Derivatives,* 2004, *18*(7), 546–550.
72. Samarth, R.M., Samarth, M. Protection against radiation-induced testicular damage in Swiss albino mice by Mentha piperita (Linn.). *Basic & Clinical Pharmacology & Toxicology,* 2009, *104*(4), 329–334.
73. Ohara, A., Matsuhisa, T. Anti-tumor promoting activities of edible plants against okadaic acid. *Food Science and Technology Research,* 2002, *8*(2), 158–161.
74. McKay, D.L., Blumberg, J.B. A review of the bioactivity and potential health benefits of peppermint tea (Mentha piperita L.). *Phytotherapy Research: An International Journal Devoted to Pharmacological and Toxicological Evaluation of Natural Product Derivatives,* 2006, *20*(8), 619–633.
75. Sim, E., Walters, K., Boukouvala, S. Arylamine N-acetyltransferases: From structure to function. *Drug Metabolism Reviews,* 2008, *40*(3), 479–510.
76. Lin, J.-P., Li, Y.-C., Lin, W.-C., Hsieh, C.-L., Chung, J.-G. Effects of (-)-menthol on arylamine N-acetyltransferase activity in human liver tumor cells. *The American Journal of Chinese Medicine,* 2001, *29*(2), 321–329.
77. Leijon, S., et al. Oral thermosensing by murine trigeminal neurons: Modulation by capsaicin, menthol and mustard oil, *Journal of Physiology,* 2019, *597*(7), 2045–2061.
78. Chumpitazi, B., Kearns, G., Shulman, R. Review article: The physiological effects and safety of peppermint oil and its efficacy in irritable bowel syndrome and other functional disorders. *Alimentary Pharmacology & Therapeutics*, 2018, *47*(6), 738–752.
79. Fang, Y., et al. Inhibition of muscular nociceptive afferents via the activation of cutaneous nociceptors in a rat model of inflammatory muscle pain. *Neuroscience Bulletin*, 2020, *36*(1), 1–10.
80. Khalil, A.F., Elkatry, H.O., El Mehairy, H.F. Protective effect of peppermint and parsley leaves oils against hepatotoxicity on experimental rats. *Annals of Agricultural Sciences,* 2015, *60*(2), 353–359.
81. Maliakal, P.P., Wanwimolruk, S. Effect of herbal teas on hepatic drug metabolizing enzymes in rats. *Journal of Pharmacy and Pharmacology,* 2001, *53*(10), 1323–1329.

82. Ogaly, H.A., Eltablawy, N.A., Abd-Elsalam, R.M. Antifibrogenic influence of Mentha piperita L. essential oil against CCl_4-induced liver fibrosis in rats. *Oxidative Medicine and Cellular Longevity,* 2018, *2018*, 4039753.
83. Kennedy, D., et al. Volatile terpenes and brain function: investigation of the cognitive and mood effects of Mentha× Piperita L. essential oil with in vitro properties relevant to central nervous system function. *Nutrients,* 2018, *10*(8), 1029.
84. Hemalakshmi, V., Thejomoorthy, P., Sriram, P., Mathuram, L. Hypoglycemic and antioxidant activities of methanolic extract of Eclipta alba in experimentally induced diabetes mellitus in rats. *Tamilnadu Journal of Veterinary and Animal Sciences,* 2012, *8*(4), 215–226.
85. Figueroa-Pérez, M.G., Pérez-Ramírez, I.F., J.A. Enciso-Moreno, I.F., Gallegos-Corona, M.A., Salgado, L.M., Reynoso-Camacho, R. Diabetic nephropathy is ameliorated with peppermint (Mentha piperita) infusions prepared from salicylic acid-elicited plants. *Journal of Functional Foods,* 2018, *43*, 55–61.
86. Barbalho, S.M., et al. Metabolic profile of offspring from diabetic Wistar rats treated with Mentha piperita (peppermint). *Evidence-Based Complementary and Alternative Medicine,* vol. 2011, *2011*, 430237.
87. Badal, R.M., Badal, D., Badal, P., Khare, A., Shrivastava, J., Kumar, V. Pharmacological action of Mentha piperita on lipid profile in fructose-fed rats. *Iranian Journal of Pharmaceutical Research: IJPR,* 2011, *10*(4), 843.
88. Talbott, H.E., et al. Wound healing, fibroblast heterogeneity, and fibrosis. *Cell Stem Cell,* 2022, *29*(8), 1161–1180.
89. Bosanquet, D.C., Harding, K.G. Wound healing: potential therapeutic options. *British Journal of Dermatology*, 2022, *187*(2), 149–158.
90. Singampalli, K.L., et al., The role of an IL-10/hyaluronan axis in dermal wound healing. *Frontiers in Cell and Developmental Biology*, 2020, *8*, 636.
91. Kumar, S., Wahab, N., Warikoo, R. Bioefficacy of Mentha piperita essential oil against dengue fever mosquito Aedes aegypti L. *Asian Pacific Journal of Tropical Biomedicine,* 2011, *1*(2), 85–88.
92. Zong, L., et al. Preliminary experimental research on the mechanism of liver bile secretion stimulated by peppermint oil. *Journal of Digestive Diseases*, 2011, *12*(4), 295–301.
93. Feitosa, K., et al. Menthol and menthone associated with acetylsalicylic acid and their relation to the hepatic fibrosis in infected mice. *Frontiers in Pharmacology,* 2017, *8*, 1000.
94. Asif, M., Saleem, M., Saadullah, M., Yaseen, H.S., Al Zarzour, R. COVID-19 and therapy with essential oils having antiviral, anti-inflammatory, and immunomodulatory properties. *Inflammopharmacology,* 2020, *28*, 1153–1161.
95. Schnitzler, P., Astani, A., Reichling, J. Antiviral effects of plant-derived essential oils and pure oil components. *Lipids and Essential Oils as Antimicrobial Agents,* 2011, 239–254.
96. Kulkarni, S.A., Nagarajan, S.K., Ramesh, V., Palaniyandi, V., Selvam, S.P., Madhavan, T. Computational evaluation of major components from plant essential oils as potent inhibitors of SARS-CoV-2 spike protein. *Journal of Molecular Structure,* 2020, *1221*, 128823.

97. Eccles, R. Menthol: Effects on nasal sensation of airflow and the drive to breathe. *Current Allergy and Asthma Reports,* 2003, 3(3), 210–214.

CHAPTER 2

Abiotic Stress and Essential Oil Composition of Peppermint: An Overview

SAILEE CHOWDHURY,[1] SUBHABROTA MAJUMDAR,[2] KOYEL KAR,[1] DIPANJAN KARATI,[3] and RANA MAZUMDER[4*]

[1]BCDA College of Pharmacy and Technology, Hidaypur, Barasat, Kolkata, West Bengal, India

[2]Calcutta Institute of Pharmaceutical Technology & Allied Health Sciences, Uluberia, Howrah, West Bengal, India

[3]School of Pharmacy, Techno India University, West Bengal, Kolkata, India

[4]Bengal College of Pharmaceutical Technology, Dubrajpur, Birbhum, West Bengal, India

*Corresponding author

ABSTRACT

The essential oil of peppermint, which is derived from plants, has a significant commercial value in perfumes, cosmetics, and medicine. Enhancing yield and preserving oil quality is crucial from a business perspective. Different biotic and abiotic elements that the plants are exposed to during their growth affect how much oil the plants produce. Because of both natural and artificial stressors, plants are subjected to varying degrees of stress. The production and quality of the bioactive components found in the oil are significantly impacted by salinization, drought, and the presence of heavy metals in the substratum. The degree and kind of stress have negative effects

Peppermint: Novel Strategies for Performance Improvement. Bilal Amhad, Tariq Ahmad Dar, and Khalid Rehman Hakeem (Eds.)

DOI: 10.1201/9781779641557-2

on the growth and development of many plants. The research conducted on certain typical abiotic stresses to which the essential oil of peppermint plants is exposed during their growth cycle and their impact on the quality and yield of oil is described in this chapter, with varying plant responses, yields, and qualities. Understanding the role played by abiotic stress may give considerable benefits to essential oil farmers and the processing sector. Improving peppermint oil (essential oil) productivity is a big task. A scientific analysis of the information on numerous significant but obscured essential oil plants will also aid in reducing the negative effects of abiotic stress.

2.1 INTRODUCTION

Medicinal plants have been used for their potent compounds to cure various diseases since the beginning of time on Earth [1]. In India, recent studies have revealed that 20,000 medicinal plant species have been documented out of 800 that can be used as phytochemicals to treat illness [2]. The medicinal and aromatic plants (MAPs) are a broad collection of species of plants that can be used in the production of medicines with no adverse effects. Medicinal plants include medicinal or medicinal reasons; the aromatic plants contain aromatic compounds that allow the plant to be used as flavoring agents and to manufacture cosmetics [3]. About 70,000 plants were used for medicinal reasons across various nations. Some plants, like mint, turmeric, garlic, and fennel are used directly as a remedy in a variety of countries all over the world. In India, the diversity of all plants comprises 2.5% of essential oil (EO) plants. It is also composed of secondary metabolites, which can also be used as pharmaceutical compounds or cosmetics. They also serve as flavoring agents and cosmetics [4].

These days, plants have been used to treat ailments, but they also provide a safe potential for a natural alternative to medicine with fewer side effects. The plant also represents an economy that goes beyond the production of pharmaceuticals. In the rural or village regions, people are also dependent upon these [2].

2.2 A BRIEF IDEA OF PEPPERMINT AND ITS ESSENTIAL OIL COMPOSITION

Essential oils (EOs) are liquid mixes composed of volatile oils that are usually extracted by steam distillation with aromatic aromas. The effects of EOs are anti-inflammatory, antiviral, and antioxidant [5–8]. Despite the huge chemical diversity, many EOs are made up of substances that are part of two families of structural compounds, specifically terpenoids and phenylpropanoids. The two families comprise numerous phenolic substances, which have antioxidant activity [9, 10]. Normally, phytochemicals work as antioxidants *via* a range of mechanisms of operation, including neutralizing free radicals and the chelation of redox-active metals, increasing endogenous antioxidant defenses, and plummeting autoxidation [11–13] (Table 2.1).

Mentha is an aromatic plant that belongs to the Lamiaceae family of plants [14]. The peppermints, as well as two varieties of spearmints, Scotch spearmint and Native spearmint, belong to the major crops that contribute to EO production across the globe and are extensively utilized as flavorings in foods, toothpaste, cosmetics, pharmaceuticals, and other products [15, 16]. Because of the high concentration of phenolic materials [17], the extracts of mentha plants could be natural antioxidants [18]. However, the majority of studies looking at the antioxidant properties of mint EOs are based on experiments using chemical substances. However, the efficacy of mint EOs in preventing oxidative stress on a cellular scale or within living organisms is not yet clearly understood. The antioxidant activity of mint EOs, in general, with regard to iron reduction as well as free radical scavenging, has been studied using chemical assays.

TABLE 2.1 Composition of Essential Oil in Peppermint

Sl. No.	Compounds	Essential oil (%) in Peppermint
1	Alpha-pinene	0.56
2	1-Octen-3-ol	0.11
3	Sabinene	0.43
4	Beta-pinene	0.81
5	3-Octanol	0.26
6	Beta-myrcene	0.20
7	Limonene	1.58
8	Linalool	0.37
9	Menthone	21.8
10	Menthol	38.45
11	Piperitone	0.65
12	Pulegone	0.91
13	Thymol	0.10
14	Eugenol	0.05
15	Germacrene D	3.24
16	Viridiflorol	0.8

2.3 STRESS ON PLANTS

2.3.1 DEFINITION OF STRESS

Plant stress refers to environmental conditions that can influence growth, development, and production. Stress can affect plants in many ways, changing their gene expression and the growth rate, cellular metabolism, and yields for crops [19]. It can be defined in three ways:

1. Physically, it is defined as the force operating on a material per unit area that causes strain and results in dimensional change is known as stress. It is typically applied to biological systems in a broader sense to characterize the effects of unfavorable forces [20].
2. Biologically, stress can refer to any situation that might harm specific organisms, entire populations, or entire communities. Another definition of stress is the overwhelming pressure that interferes with a person's ability to carry out their daily activities or the circumstances that hinder plants from developing to their full genetic potential.
3. Agriculturally, stress lowers crop productivity or depletes biomass.

2.3.2 CLASSIFICATION OF STRESS

Ecologists, physiologists, and agronomists generally classify challenges to plants into two types: biotic stress and abiotic stress. Abiotic stress is a result of interactions between the organisms and their physical environment. On the other hand, the biotic stressors result from the interaction between different species. Biological stressors, in their simplest forms, can be caused by species-specific competition for resources, such as predation, parasitism, and predation, and the effect of allelopathic chemicals produced by one species that affect another.

2.3.2.1 ABIOTIC STRESS

One of the major challenges for agriculture is managing abiotic stress. Stress from abiotic causes can permanently limit the available crop varieties and the amount of agricultural production across large regions, and severe episodes may result in the complete failure of crops. Abiotic stress causes harm to the national economy, food security, and individual farmers' and their families' well-being [21].

2.3.2.2 BIOTIC STRESS

Living creatures, including bugs, viruses, fungi, nematodes, arachnids, insects, and weeds, can cause stress to the biotic system in plants. The death of plants can be caused by the organisms that cause biotic stress, directly denying their hosts the nutrients they need. Post- and preharvest losses are a major cause of biochemical stress. Plants can survive biotic stresses but lack an adaptive immune system responding to specific advanced techniques. The genome of plants, red in these cells, governs the mechanisms for defense to respond to stresses. The plant's genome contains many genes immune to various biochemical stressors. Crop plants suffer abiotic stresses due to living factors like sunshine, salinity, heat, cold, floods, and drought, which are different from stress caused by biotics. The kind of biotic strain that can be put on crops is contingent on the conditions where they are grown and their capacity to endure that specific type of stress. Since chewing insects decrease leaves' area, and viral infections reduce the efficiency of photosynthesis for each leaf surface, numerous environmental stressors can significantly affect photosynthesis (Figure 2.1) [22].

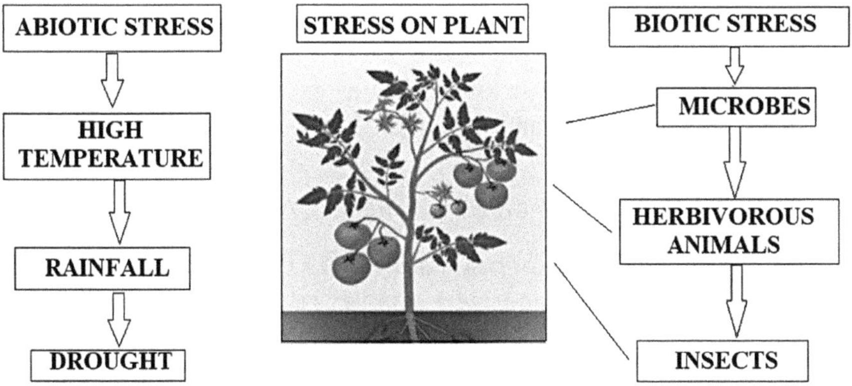

FIGURE 2.1 Common plant stresses.

2.3.3 A BRIEF IDEA ABOUT ABIOTIC STRESS

2.3.3.1 DEFINITION

The potential harms of drought, salt, flooding, metal toxicity, low nutrient levels, and high and low temperatures are a few examples of stressors that can be attributed to abiotic causes. Abiotic stressors can include very localized, intermittently occurring circumstances such as shade and UV exposure, pollution from the air, photo inhibition hail, wind, and gaseous insufficiency. Stress from abiotic causes can affect plants as a result of the absence of a resource that is essential or due to the presence of a substance that is harmful or poisonous at excessive levels. Water availability can also become a source of stress, which could be caused by excessive (flooding) and inadequate (drought). Stressors from biotic and abiotic sources can be interspersed in real life [23]. Abiotic stressors negatively impact the development, yield, seed, and overall quality of the crops. Crops are affected by drought (water stress), excessive watering (waterlogging), and extreme temperatures (cold or frost and temperatures), as well as salinity and mineral toxicity. There is a possibility of future shortages of fresh water, which will cause an increase in levels of abiotic stress [24]. To ensure health and security over the coming years, there is a need to find varieties of plants that can withstand abiotic stressors. The roots of plants are the primary protection against the effects of abiotic stress. If the soil on which the plant grows is robust and diverse in its biological composition, it stands the best chance to survive difficult circumstances. The disruption of the Na^+/K^+ ratio within the cell's cytoplasm in plants is one of the major reactions to environmental stressors like high salinity. Abscisic acid, which is the phytohormone that plants use, is essential for adapting to environmental stressors such as excessive salt, dehydration, low temperatures, or mechanical damage [25].

2.3.3.2 THE IMPACTS OF ABIOTIC STRESS ON PLANTS

An array of biotic and abiotic stresses creates the conditions the plants are naturally in. Stresses can trigger complex responses in plants, too. To identify the regulatory hubs embedded in complex systems, techniques from systems biology allow for a multi targeted approach. Systems biology seeks to integrate the molecular components of an organism, such as their proteins, metabolites, and transcripts, into functioning systems or models designed to predict and capture the constantly changing activities of an organism within a range of settings.

Animals have come up with methods to reduce the impact of the variances, but plants cannot accomplish this because of their sessile growth habits. Plants depend on their internal systems to adapt to changes in the external world. When the environment is in flux, the plants must adapt to their environment to function. Healthy internal changes help balance typical external fluctuations without affecting growth or development [26]. Abiotic stress can be the cause that a plant may suffer physical harm as a result of persistently adverse circumstances. The term "biotic stress" refers to the adverse effect of nonliving objects upon living creatures within a static setting. Nonliving factors must alter the surrounding environment outside of its normal fluctuations to dramatically impact the individual's output and demographics or physiology of the living thing. Stress on the physical environment has been identified as a significant threat to food security as a result of ongoing changes in the climate and environmental degradation caused by human activities. A crop's yield is severely impacted by environmental issues like salinity stress, water scarcity tension, nutritional imbalances (including mineral toxicity and deficiency), and extreme temperatures [27]. The impacts of heavy metals, salt, drought, and cold on the growth of plants and their agricultural production are significant. The drought causes abiotic stress that impacts plants at the molecular and morphological scales and manifests throughout growth. Drought affects three key stages of the plant: vegetative development, pre-anthesis, and the terminal period [28]. The terminal phase is the most severe. Plants' physiological reactions to stress are leaf withering and abscission. They also experience a reduction in leaf size and a decrease in transpirational water loss [29].

Development of the crop greatly helps in addressing the issue of overwater consumption in agricultural practices in situations of drought. One of the subtlest physiological methods to promote the development of cells is a reduction in the pressure of turgor. It is a problem for water flow between the xylem and the adjacent elongating cells of the higher plant tissues during dry conditions, resulting in the elongation of cells. Furthermore, drought stresses cause cells to elongate, mitosis damage, and reduce crop growth, leaf size, plant height, and yield [30]. Mechanisms for avoiding escape, as well as tolerance, can be component of abiotic stress resistance. The modulation of osmotic pressure can reduce the adverse impacts of stress and aid in the buildup of active liquids within the cytoplasm to maintain the cell's water balance [31].

Also, low temperatures can impact the longevity and expansion of plant species. In many cases, cold stress causes large losses in crop production because of a decrease in the growth of plants and decreased efficiency [32]. It hinders the development of plants and yield by causing tissue and cell dehydration along with cellular crystallization. The increase in electrolyte leakage when temperatures are low is caused by a decrease in conductivity in the membrane, increased water viscosity, and a hampered closing of the hydroactive stomata. Furthermore, it blocks metabolic processes, uses energy, and triggers the production of free radicals, a consequence of oxidative stress [33].

It is a fact that 38% of the planet's people live in areas where it is not rainy or less rainfall, while approximately 45% of the world's agriculture is affected by saltwater, covering an area of more than 3106 km^2. Roughly 6% of the globe's soil surface is impacted by this issue [34].

Plants' most significant abiotic stress throughout their development and growth is the water-related constraint caused by lack of rainfall, cold temperatures, and salinity. Abiotic stressors are connected and significantly influence the impact of water on plants at both the systemic and cellular levels. The changes in morphology, biochemistry, physiology, and molecular structures have detrimental effects on the growth of plants and their development.

2.3.3.2.1 The Effect of Drought on Plants

Drought stress affects every phase of the plant's phonological development. The effects range between morphological and molecular levels. Many biological processes in plants affect yield and are susceptible to drought [33]. Based on the severity and length of drought, a variety of species of plants can suffer declines in their yield; however, the strain of drought following anthesis can be detrimental to the production of grain regardless of the intensity of the drought [35]. A drought-related stress can limit flowering and grain filling. This impacts the grain's quality and amount. Nitrogen (N), phosphorus (P), and potassium (K), as well as other macro- and micronutrients, are vital for the growth of plants. Even though drought stresses increase N and significantly decrease P, K has no evident effects.

2.3.3.2.2 Effect of Cold on the Plants

The severe abiotic stress, commonly called cold stress, could reduce the yield of crops and create surface lesions. This can affect their development and growth. The cells' structure, functioning membranes, and different metabolic processes can trigger these signs [36]. According to research, the environment often exhibits an increase in extreme temperature frequency, intensity, and duration. Increased temperature fluctuations are caused by climate change and result in a more frequent occurrence of cold weather. The plants susceptible to extreme cold have lower development rates, fewer uses of the most valuable varieties, and less yield. The plants use various biological, metabolic, and molecular modifications in stressful conditions. The techniques are first able to produce changes to help protect the plant and, later, cold adaptation, which improves the ability of plants to handle extreme cold [37].

2.3.3.2.3 The Effect of Heat on the Plant

The stress of high temperatures also hinders crop growth and development because of the succession of biochemical, morphological, and physiological modifications. These are among the primary factors that hinder the productivity of crops globally because of the increase in air temperatures. The growth phases and spread of the agricultural plant could change because of the increasing temperature [38]. The deterioration of proteins, the interruption of protein synthesis, the activation of vital enzymes, and damaged membranes are all due to high-temperature stress. The stress of high temperatures could affect the way cells divide. These negative impacts are likely to hinder plant growth and cause oxidative injuries. In addition, a brief exposure to extreme temperatures when filling seeds can lead to faster filling, which reduces production and quality.

2.3.3.2.4 The Effect of Salt on Plants

In the wake of natural and agriculture-related occurrences like irrigation, the quantity of land that is affected by high salinity increases across the globe. The effects of salinity are twofold on plant growth, namely, ionic stress and osmotic stress. Oxidative stress can also be seen due to this. The negative effects of salinity alter a variety of metabolic and physiological processes within plants. Many symptoms, such as decreased leaf size, increased leaf size, silent leaf abscission root-shoot necrosis, and smaller internodes, are often accompanied by reactions to these changes. When exposed for a long time to salinity stress, it slows the growth and increases the rate of cell senesces. The most significant harm, and one that may cause additional symptoms, is a decrease in growth. Salt shocks can result in the death of programmed cells [39].

2.3.3.2.5 Effect of Heavy Metal

Release of high-risk metals in the air due to the activities of humans or natural processes is an extremely widespread problem. These potentially harmful elements are called heavy metals, also known as trace metals. The traces of components found in soils are related to trace metals. Heavy metals exist in soil in various forms due to their varying solubility/bioavailability. Geochemical properties of heavy metals found in the soil, in plant uptake, and agricultural production depend on a range of soil physical properties. Many physiological, biochemical, and morphological functions in plant tissues can be negatively affected by the deposition of heavy metals in the plant tissues directly or indirectly, affecting crop production. The effects of heavy metals on crop production are mediated by affecting a range of plant physiological functions such as seed germination, the accumulation and storage of seeds during the growth of plants, germination, and photosynthesis [40].

By the reaction of oxygen species, disturbance of the redox balance, and oxidative stress, toxic metals on the cell platform can reduce plant efficiency. Metals are mainly introduced into plants through the roots of the soil (Figure 2.2) [41].

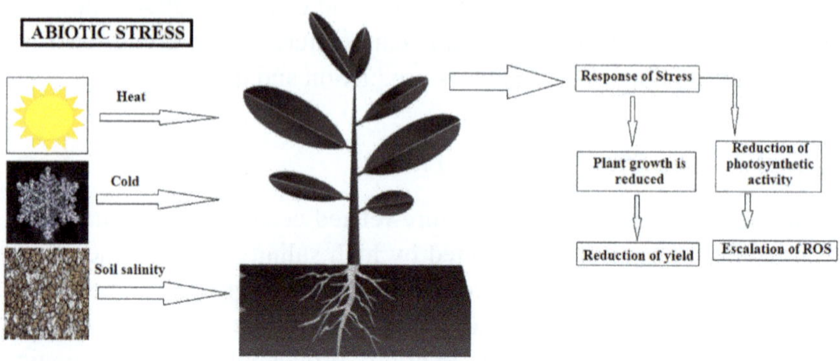

FIGURE 2.2 Different types of abiotic stress factors in plants.

2.3.3.3 MECHANISMS OF DEFENSE USED BY PLANTS TO FIGHT THE ABIOTIC EFFECTS OF STRESS

They are famous for establishing innovative mechanisms in their metabolism to mitigate the detrimental effects induced by external environmental factors, since abiotic stressors are ones that cannot be maintained and decreased by plants outside. The plants employ various strategies to counteract these environmental stressors [42]. The steps that plants take gradually to reduce the negative effects of stress caused by antibiotics are examined. The graph shows that the plants typically produce a high quantity of ROS during an abiotic stress. Oxygen antioxidants increase the osmolyte level, and the activation of stress-responsive genes is the subsequent three steps that occur because of excessive production. As the genes for stress response are activated, plants have a higher tolerance and can withstand these dangerous temperatures. Identifying targeted genes is crucial since they are the only genes contributing to the overall system. Therefore, it is possible to conclude that reducing oxidative stress improves stress tolerance parameters and longevity of plants operating under these conditions [43]. The oxidative stress reduction process is crucial to ensure the long-term viability of plants.

2.3.3.3.1 The Steps Involved in Overcoming Abiotic Stress in Plants

The capacity of plants to create excess ROS when undergoing an oxidative stress reduction reaction to environmental challenges can increase the stress tolerance and long-term viability of plants (Figure 2.3) [44].

2.4 EFFECT OF DIFFERENT TYPES OF ABIOTIC STRESS ON ESSENTIAL OIL COMPOSITION OF PEPPERMINT

Abiotic stress can be defined as the adverse impact that nonliving elements have on organisms within a particular environment. These stresses affect the plant's medicinal as well as other properties. The main reason for abiotic stress is the unintentional change in the environment. The environmental change can be prolonged drought, heavy rains, flooding, increased soil salinity and heat waves, and even frost damage. Worldwide, the main

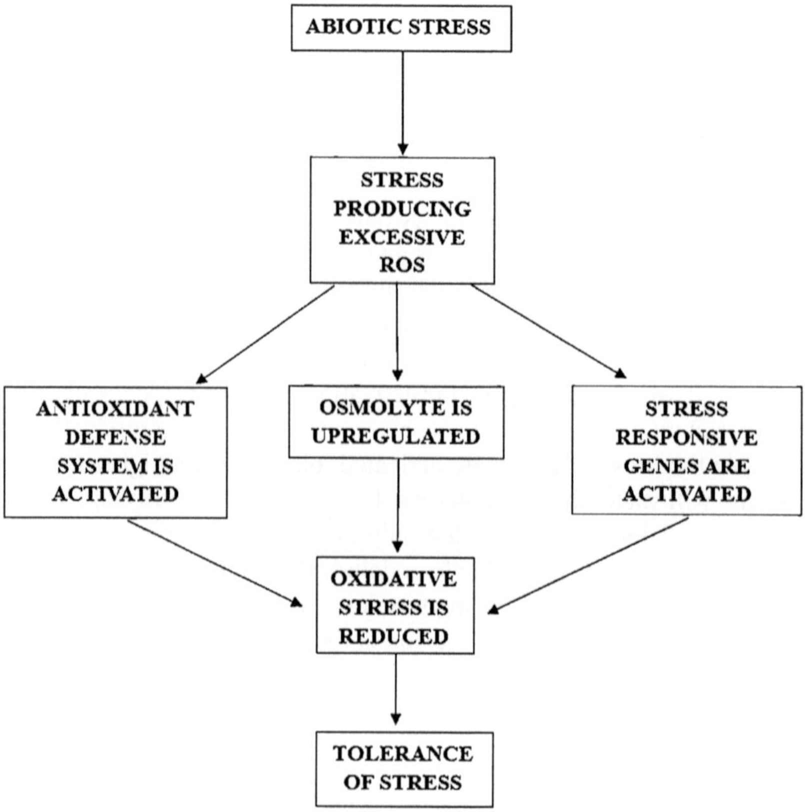

FIGURE 2.3 Steps involved in overcoming abiotic stress by plants.

causes of crop loss are drought, flooding, and soil salinity/sodicity. These factors negatively affect not only plant growth but also productivity. Many parts of the globe are affected by salt stress [3]. The abiotic stress has a negative effect on the plants because it alters the metabolism of the cells by producing an excess of reactive oxygen species in the plant. The reactive oxygen species in plants have a double role. They are both toxic and key regulators of many biological processes, such as growth, programmed death, cell cycles, hormone signals, and cell response and development. Primary metabolites are responsible for secondary metabolites in most higher plants. The concentrations of different secondary plant products depend on growing conditions. Secondary metabolite is used in many applications, including nutritive and medicinal products, food additives, flavors, pharmaceuticals, industrial pharmaceuticals, as well as pharmaceuticals.

Mentha odora Salisb. are perennials belonging to the Lamiaceae family, which originated in Europe. They have spread all over the world. *M. odora* Salisb. is a hexaploid medicine plant that has been attributed to a hybrid between *M. spicata* L. and *M. aquatica* L. Both *Mentha* species have a high level of consumption around the globe and are important for their EO production. The EO is composed of menthone, isomenthone, 1,8-cineole, menthyl acetate, menthofuran, limonene, β-myrcene, β-caryophyllene, carvone, and pulegone. *M. piperita* has two pathways that produce monoterpenes or sesquiterpenes. These are important secondary metabolites in volatile oils. Temperature, salt content, drought conditions, anaerobic or flooding conditions, and other abiotic stressors can affect productivity and growth.

In the last decade, there has been a tremendous rise in demand and consumption of medicinal plants. Aromatic plants, their EOs, flavors, and fragrances are used in industry to contribute to the economy of developing nations through exports. The use of chemotherapeutics is important in the management and treatment of clinical conditions.

It is possible to produce large amounts of chemotherapeutics, such as cancer drugs, diabetes, thyroid hormones, and malaria agents, from plants that are grown under abiotic conditions. At present, research is being conducted to identify stress factors that EO-producing plants face. Stress plays a major role in the growth and development process of EOs. Understanding this will help to provide resources and insight into the problem. EOs are also important in the development of resistance to biotic stress. It is also important to investigate their role in the plant's defense mechanisms. The ability of plants to adapt and survive under stressful conditions is affected by changes in the substratum. A variety of species have a number of different factors to consider. So far, only a few plants have been studied in depth, mainly Lamiaceae. In the future, since novel components and plants for EO are being purified and isolated, it is important to focus on less-known EO species (Figure 2.4).

2.4.1 WATER STRESSES

The plants are also affected by water stress and the oils that they contain. Water stress reduces plant quality, height, and area. Water stress has

FIGURE 2.4 Abiotic conditions for the peppermint plant.

been shown to affect the composition of EOs by changing the levels of constituents depending on the plant. The effects of water stress on the compositions of EOs distilled out from cumin seeds (*Cuminum cyminum*) were observed [3].

2.4.2 STRESS ON SALINITY

The degradation of salt is the main cause of salinization. The land is affected as it decomposes nutrients and decreases soil quality. The primary carbon metabolism and plant growth are affected by the ion-toxic effect of salinity (Table 2.2) [45].

TABLE 2.2 Some Examples of Effect of Salinity on Increase/Decrease of Essential Oil [45]

Serial No	Plant Name	Salinity Level (mM)	Effect on the quantity of Essential Oil
1	*Mentha piperita*	0–150	Decrease
2	*Salvia sclarea*	0–75	Decrease
3	*Cymbopogon schoenanthus*	0–150	Increase
4	*Coriandrum sativum*	0–75	Increase
5	*Matricaria chamomile*	0–336	Increase

An experiment in pot culture was performed to determine the effects of salinity and EO on *Mentha pulegium*. Menthone (51%) was the most common compound isolated from *M. pulegium* plants grown under control conditions. Pulegone (23.9%) and neomenthol (15.3%) were next. Pulegone was found to increase by 60% in plants treated with salt. Under salinity, the composition changed from menthone under controlled conditions to become pulegone.

2.4.3 TEMPERATURE STRESS

The temperature is a major biotic factor that affects the productivity of plants. High temperature and low temperature can lead to a variety of physiological, biochemical, and molecular changes. Heat stress causes leaf senescence and cell membrane damage. It also leads to a decrease in chlorophyll levels, as well as denaturation of proteins. Several studies have been conducted on the effects of temperature on MAP growth, as well as EO quantity and quality. Al-Huqail et al. Studied the effect of temperature stress on sweet basil (*Ocimum basilicum*). Al-Huqail et al. observed a decrease in the growth of plants under high temperatures (35°C, 45°C, 55°C) when compared to the 25°C control. This could be because high temperature affects plant metabolism. Similarly, plant growth parameters (flower yield and plant height) for chamomile (*Matricaria chamomilla*) decreased at high temperatures (15°C, 20°C, and 25°C) while these parameters increased at a low temperature (12°C). Fennel (*Foeniculum vulgare*), on the other hand, showed a reduction in fresh and dried weight under cold stress. A second study found that an increase in temperature could cause American ginseng to senesce earlier and reduce photosynthetic rates. A decline in total biomass was also observed. Heat stress has a significant impact on the composition of EOs from *M. arvensis* and *M. piperita*. The percentage of menthol decreases for both mint species as the temperature increases, whereas the percentages of pulegone or menthyl acetate increase. Menthone, menthofuran, and their percentages also vary greatly with temperature. The antimicrobial activity of EOs from two species of mint was analyzed. The oils tested showed strong antimicrobial properties, dose-dependent, and directed at pathogenic strains that are resistant to antimicrobial agents [45].

2.5 CONCLUSION

There were significant qualitative and quantitative differences in the chemical compositions between *M. pipperita* EO obtained from plants that had been grown with different irrigation treatments. In all stages of growth examined, mild stress increased EO concentration while moderate stress reduced it. The EO obtained from the peppermint plant in this study did not meet acceptable quality standards and was not produced in sufficient quantities to prove that peppermint could be successfully grown under different stress conditions. Recommended commercial production of peppermint without stress from water.

KEYWORDS

- **adenomyosis**
- **endometriosis**
- **gynecologic cancer**
- **gynecologic robotic surgery**
- **myomectomy**
- **sacrocolpopexy**
- **uterine bleeding**
- **uterine prolapse**

REFERENCES

1. Fazili, M.A., Masood, A., Wani, A.H., Khan, N.A. Essential oil of mint: Current understanding and future prospects. *Biodiversity and Biomedicine*, 2020, *1*, 293–304.
2. Aftab, T. A review of medicinal and aromatic plants and their secondary metabolites status under abiotic stress. *Journal of Medicinal Plants*, 2019, *7*(3), 99–106.
3. Israel, A., Langrand, J., Fontaine, J., Lounès-HadjSahraoui, A., Significance of arbuscular mycorrhizal fungi in mitigating abiotic environmental stress in medicinal and aromatic plants: A review. *Foods*, 2022, *11*(17), 2591.
4. Biswas, S., Koul, M., Bhatnagar, A.K. Effect of salt, drought and metal stress on essential oil yield and quality in plants. *Natural Product Communications*, 2011, *6*(10), 1934578X1100601036.
5. Baydar, N.G., Özkan, G., Sa˘ gdiç, O. Total phenolic contents and antibacterial activities of grape (Vitis vinifera L.) extracts. *Food Control*, 2004, *15*, 335–339.
6. Sökmen, M., Serkedjieva, J., Daferera, D., Gulluce, M., Polissiou, M., Tepe, B., Akpulat, H.A., Sahin, F., Sokmen, A. In vitro antioxidant, antimicrobial, and antiviral activities of the essential oil and various extracts from herbal parts and callus cultures of Origanum acutidens. *Journal of Agricultural and Food Chemistry*, 2004, *52*, 3309–3312.
7. Dundar, E., Olgun, E.G., Isiksoy, S., Kurkcuoglu, M., Baser, K.H.C., Bal, C. The effects of intra-rectal and intra-peritoneal application of Origanum onites L. essential oil on 2, 4, 6-trinitrobenzenesulfonic acid-induced colitis in the rat. *Experimental and Toxicologic Pathology*, 2008, *59*, 399–408.
8. Liu, Y., Song, M., Che, T., Bravo, D., Pettigrew, J. Anti-inflammatory effects of several plant extracts on porcine alveolar macrophages in vitro. *Journal of Animal Science*, 2012, *90*, 2774–2783.
9. Ruberto, G., Baratta, M.T. Antioxidant activity of selected essential oil components in two lipid model systems. *Food Chemistry*, 2000, *69*, 167–174.
10. Djeridane, A., Yousfi, M., Nadjemi, B., Boutassouna, D., Stocker, P., Vidal, N. Antioxidant activity of some Algerian medicinal plants extracts containing phenolic compounds. *Food Chemistry*, 2006, *97*, 654–660.
11. Lobo, V., Patil, A., Phatak, A., Chandra, N. Free radicals, antioxidants and functional foods: Impact on human health. *Pharmacognosy Reviews*, 2010, *4*, 118.
12. Oh, J., Jo, H., Cho, A.R., Kim, S.-J., Han, J. Antioxidant and antimicrobial activities of various leafy herbal teas. *Food Control*, 2013, *31*, 403–409.
13. Arzani, A., Zeinali, H., Razmjo, K. Iron and magnesium concentrations of mint accessions (Mentha spp.). *Plant Physiology and Biochemistry*, 2007, *45*, 323–329.
14. Bensabah, F., Houbairi, S., Essahli, M., Lamiri, A., Naja, J. Chemical composition and inhibitory effect of the essential oil from mentha spicata irrigated by wastewater on the corrosion of aluminum in 1 molar hydrochloric acid. *Portugaliae Electrochimica Acta*, 2013, *31*, 195–206.
15. Riachi, L.G., De Maria, C.A. Peppermint antioxidants revisited. *Food Chemistry*, 2015, *176*, 72–81.
16. Naidu, J.R., Ismail, R., Yeng, C., Sasidharan, S., Kumar, P. Chemical composition and antioxidant activity of the crude methanolic extracts of Mentha spicata. *Journal of Phytological*, 2012, *4*, 13–18.

17. Dorman, H.J., Kosar, M., Kahlos, K., Holm, Y., Hiltunen, R. Antioxidant properties and composition of aqueous extracts from Mentha species, hybrids, varieties, and cultivars. *Journal of Agricultural and Food Chemistry*, 2003, *51*, 4563–4569.
18. Zheljazkov, V.D., Cantrell, C.L., Astatkie, T., Hristov, A. Yield, content, and composition of peppermint and spearmints as a function of harvesting time and drying. *Journal of Agricultural and Food Chemistry*, 2010, *58*, 11400–11407.
19. Dresselhaus, T., Hückelhoven, R. Biotic and abiotic stress responses in crop plants. *Agronomy*, 2018, *8*(11), 267.
20. Bechtold, U., Field, B. Molecular mechanisms controlling plant growth during abiotic stress. *Journal of Experimental Botany*, 2018, *69*(11), 2753–2758.
21. Zhang, H., Zhu, J., Gong, Z., Zhu, J.K. Abiotic stress responses in plants. *Nature Reviews Genetics*, 2021, 1–16.
22. Gull, A., Lone, A.A., Wani, N.U.I. Biotic and abiotic stresses in plants. In: *Abiotic and Biotic Stress in Plants*, 2019; pp. 1–19.
23. Corpas, F.J. Hydrogen sulfide: A new warrior against abiotic stress. *Trends in Plant Science*, 2019, *24*(11), 983–988.
24. Zandalinas, S.I., Fichman, Y., Devireddy, A.R., Sengupta, S., Azad, R.K., Mittler, R. Systemic signaling during abiotic stress combination in plants. *Proceedings of the National Academy of Sciences*, 2020, *117*(24), 13810–13820.
25. Nadarajah, K.K. ROS homeostasis in abiotic stress tolerance in plants. *International Journal of Molecular Sciences*, 2020, *21*(15), 5208.
26. Tandzi, L.N., Bradley, G., Mutengwa, C. Morphological responses of maize to drought, heat, and combined stresses at the seedling stage. *Journal of Biological Sciences*, 2019, *19*(1), 7–16.
27. Fahad, S., Bajwa, A.A., Nazir, U., Anjum, S.A., Farooq, A., Zohaib, A., et al. Crop production under drought and heat stress: Plant responses and management options. *Frontiers in Plant Science*, 2017, *29*, 8.
28. Shavrukov, Y., Kurishbayev , A., Jatayev, S., Shvidchenko, V., Zotova, L., Koekemoer, F., et al. Early flowering as a drought escape mechanism in plants: How can it aid wheat production? *Frontiers in Plant Science*, 2017, *17*, 8.
29. Fghire, R., Anaya, F., Ali, O.I., Benlhabib, O., Ragab, R., Wahbi, S. Physiological and photosynthetic response of quinoa to drought stress. *Chilean Journal of Agricultural Research*, 2015, *75*(2), 174–183.
30. Mukamuhirwa, A., Persson Hovmalm, H., Bolinsson, H., Ortiz, R., Nyamangyoku, O., Johansson, E. Concurrent drought and temperature stress in rice-A possible result of the predicted climate change: Effects on yield attributes, eating characteristics, and health-promoting compounds. *International Journal of Environmental Research and Public Health*, 2019, *16*(6), 104.\
31. Yoshida, T., Fujita, Y., Sayama, H., Maruyama, K., Mizoi, J., Shinozaki, K., Yamaguchi-Shinozaki, K. AREB1, AREB2, and ABF3 are master transcription factors that cooperatively regulate ABRE-dependent ABA signaling involved in drought stress tolerance and require ABA for full activation. *Plant Journal*, 2010, *61*(4), 672–685.
32. Browse, J., Xin, Z. Temperature sensing and cold acclimation. *Current Opinion in Plant Biology*, 2001, *4*(3), 241–24.

33. Farooq, M., Wahid, A., Kobayashi, N., Fujita, D., Basra, S.M.A. Plant drought stress: Effects, mechanisms, and management. *Agronomy for Sustainable Development*, 2009, *29*(1), 185–212.
34. Volkov, V. Salinity tolerance in plants. Quantitative approach to ion transport starting from halophytes and stepping to genetic and protein engineering for chickpea production. *Agronomy*, 2018, *8*(8), 145.
35. Cattivelli, L., Rizza, F., Badeck, F-W., Mazzucotelli, E., Mastrangelo, A.M., Francia, E., et al. Drought tolerance improvement in crop plants: An integrated view from breeding to genomics. *Field Crops Research*, 2008, *105*(1–2), 1–14.
36. Barrero-Sicilia, C., Silvestre, S., Haslam, R.P., Michaelson, L.V. Lipid remodeling: Unravelling the response to cold stress in Arabidopsis and its extremophile relative Eutremasalsugineum. *Plant Science*, 2017, *263*, 194–200.
37. Flowers, T.J. Improving crop salt tolerance. *Journal of Experimental Botany*, 2004, *55*(396), 307–319.
38. Porter, J.R. Rising temperatures are likely to reduce crop yields. *Nature*, 2005, *436*(7048), 174–174.
39. Parida, A.K., Das, A.B. Salt tolerance and salinity effects on plants: A review. *Ecotoxicology and Environmental Safety*, 2005, *60*(3), 324–349.
40. Uzu, G., Sobanska, S., Aliouane, Y., Pradere, P., Dumat, C. Study of lead phytoavailability for atmospheric industrial micronic and sub-micronic particles in relation with lead speciation. *Environmental Pollution*, 2009, *157*(4), 1178–1185.
41. Pourrut, B., Jean, S., Silvestre, J., Pinelli, E. Lead-induced DNA damage in Vicia faba root cells: Potential involvement of oxidative stress. *Mutation Research/Genetic Toxicology and Environmental Mutagenes*is, 2011, *726*(2), 123–128.
42. Sharma, A., Shahzad, B., Rehman, A., Bhardwaj, R., Landi, M., Zheng, B. Response of the phenylpropanoid pathway and the role of polyphenols in plants under abiotic stress. *Molecules*, 2019, *24*(13), 2452
43. Park, C.J., Park, J.M. Endoplasmic reticulum plays a critical role in integrating signals generated by both biotic and abiotic stress in plants. *Frontiers in Plant Science*, 2019, *10*, 399.
44. Kaur-Sawhney, R., Tiburcio, A., Altabella, T., Galston, A. Polyamines in plants. *Journal of Cell and Molecular Biology*, 2003, *2*, 1–12.
45. Heydari, M., Zanfardino, A., Taleei, A., ShahnejatBushehri, A.A., Hadian, J., Maresca, V., Sorbo, S., Di Napoli, M., Varcamonti, M., Basile, A., Rigano, D. Effect of heat stress on yield, monoterpene content and antibacterial activity of essential oils of Mentha x piperita var. Mitcham and Mentha arvensis var. piperascens. *Molecules*, 2018, *23*(8), 1903.

CHAPTER 3

Influence of Selective Mineral Nutrients on the Growth and Yield of Peppermint

MEHTAP AYDIN

Department of Molecular Biology and Genetics, Faculty of Engineering and Natural Sciences, Istanbul Sabahattin Zaim University, Istanbul, Türkiye

ABSTRACT

The genus Mentha, commonly referred to as peppermint, is a member of the taxonomic family Lamiaceae, also known as the mint family, and is found across the world's temperate zones. Aromatic Mentha is a flower used in folk medicine to treat sickness, headaches, fever, red eyes, and sore throats. It is also used as a flavoring and in functional teas. The variety and environmental factors, such as temperature, moisture, soil nutrients, and biotic and/or abiotic stress, affect the chemical makeup of peppermint essential oils. In medicinal and aromatic plants, the availability of nutrients is critical for improving biomass yield and the quality of essential oils. In plant metabolism, essential trace elements such as iron (Fe), zinc (Zn), boron (B), and magnesium (Mg) are vital because they serve as cofactors in many enzymatic reactions and are structural components of biomolecules. These mineral components greatly enhance the therapeutic qualities and medicinal worth of peppermint medicines. Additionally, they are essential for optimizing grain productivity

3.1 INTRODUCTION

Peppermint (*Mentha piperita* L.) is a fragrant perennial herb belonging to the Lamiaceae family. Since peppermint has a large number of bioactive

Peppermint: Novel Strategies for Performance Improvement. Bilal Amhad, Tariq Ahmad Dar, and Khalid Rehman Hakeem (Eds.)

DOI: 10.1201/9781779641557-3

compounds, including menthol, menthyl acetate, and menthone, it is highly valued in aromatic and medicinal plants. It is extensively utilized for treating various health issues, including pain relief, disinfection, gas relief, congestion relief, digestive aid, and as a stimulant, among other applications [1]. It is known that the oil it contains has many properties, such as antimicrobial, antibacterial, antifungal, antiviral, and antioxidant [2]. The temperature, humidity, abiotic and biotic stress, and soil nutrients all have an impact on the chemical structure of peppermint essential oils. As a result, improving the quality of essential oils and biomass yield in medicinal plants is significantly influenced by soil nutrients [3].

Plants need 16 essential nutrients, including micronutrients and macronutrients, to complete their life cycle and achieve productivity [4]. The soil supplies macronutrients (nitrogen, phosphorus, potassium, calcium, magnesium, and sulfur) in significant amounts. Along with these micronutrients, the soil also provides comparatively lesser amounts of iron, manganese, boron, molybdenum, copper, zinc, chlorine, and cobalt. It is important to have an adequate supply of nutrients and maintain the correct proportions. Throughout their lifespan, living organisms need micronutrients in limited amounts to regulate various physiological processes [5]. Certain physical and biochemical processes involved in plant germination and seedling production are influenced by micronutrients. The plant's ability to withstand abiotic or biotic stress is dependent on the existence of adequate levels of micronutrients, including zinc, boron, and magnesium. It provides the development of seeds and seedlings, especially in the germination period [5, 6]. Necessary mineral elements, like nitrogen (N), carbon (C), and phosphorus (P), are needed in relatively larger quantities since they form the building blocks for essential biomolecules, including amino acids, nucleotides, and carbohydrates. Moreover, they play a crucial role in maximizing grain yield [6].

3.2 IRON

Iron, which is among the micronutrients that plants need most, is a part of the catalytic group and is necessary for chlorophyll synthesis [5]. Even though there is a lot of iron in the crust of the Earth, plants find it difficult to absorb it, which causes iron deficits in plants that are malnourished [7]. Between pH 7 and pH 8.5, inorganic iron is not very soluble.

Generally speaking, especially in calcareous soils, the concentration of iron species in soil solutions is less than 10% of what is needed for sufficient plant feeding. Iron is essential for many different types of organisms' functions, including redox reactions, photosynthesis, nitrogen metabolism, and symbiotic nitrogen fixation [8, 9]. Iron deficiency inhibits chlorophyll biosynthesis, causing the plant to grow slowly and reduce its nutritional value [10].

3.3 NITROGEN

One of the elements that is most vital to plant metabolism is nitrogen (N). It has an essential place in all vital cycles and metabolism in plants. It is responsible for the continuation of critical processes in the plant and is directly related to proteins, which are an essential structural component. Therefore, nitrogen is an indispensable nutrient for crop production. Nitrogen increases food quality while also increasing production [11]. Enhancing photosynthetic activities, leaf area output, leaf area longevity, and net assimilation rate are all possible with the right amount of nitrogen (N). For a crop to yield more, it is essential to maximize the leaf area and total leaf biomass [12].

It was shown in a study conducted in 2015 that nitrogen improves the photosynthetic rate and osmotic balance of the plant by regulating the production of proline and ethylene. It has been determined that it reduces the effects of drought stress by synthesizing enzymes and starch in the leaf system of plants under drought stress [13, 14].

3.4 BORON

A vital mineral nutrient is boron (B), which plants utilize in minute amounts. It is well recognized to be crucial for the metabolism of proteins and carbohydrates, formation and synthesis of cell walls, membrane function, and actions of phenol and indole acetic acid [15]. One of the fastest responses of the plant to boron deficiency is the cessation of cell division and the elongation of secondary roots. It is also known that boron enables the pollen tube to elongate. If the amount of boron is low, it causes the internodes to be short and stem cracks to occur [5]. In case of boron deficiency, the yield, quality, membrane integrity, cell wall structure, protein, nucleic acid, and carbohydrate integrity of the product are seriously impaired. Boron deficiency disrupts the flowering and fruiting rate of plants [16].

3.5 PHOSPHOROUS

Abiotic stress responses and plant resistance are significantly influenced by adequate P feeding. The majority of plants in agricultural areas face P deficiency. When plants are exposed to this deficiency, it can be challenging to diagnose the deficiency of this mineral because its symptoms are not immediately visible [17]. Phosphorus takes part in many cellular processes, such as protecting the plant membrane structure and synthesis of biomolecules. It also contributes to the metabolism of carbohydrates, enzyme activity, and cell division. It stimulates seed germination in the plant, ensures the development of roots, and increases the quality of crop yield. Considering these effects, phosphorus is necessary in all periods from germination to maturation [18].

3.6 POTASSIUM

One of the most prevalent nutrients in plants is potassium (K). There was a continuous range of 100–200 mM for the K concentration in the cytoplasm. K is essential in protein synthesis, phloem transport, cation–anion balance, protein synthesis, and stress resistance [19]. Drought, one of the abiotic stresses faced by plants, causes a decrease in crop productivity in plants. Both root growth and K^+ diffusion rates of the plant are restricted. The resulting low K rates reduce K absorption [19].

3.7 CALCIUM

Calcium is one of the minerals that acts as a messenger within the cell, and the growth and development of the plant are aided by it. It maintains the structural integrity of the cell membrane and mediates ROS concentration [20]. As a cation that protects the vacuole from both inorganic and organic anions, calcium serves as an intracellular messenger by acting as divalent Ca^{2+} for membrane structural role-playing. The xylem transports calcium from the soil solution to the roots and then to the shoot [21].

3.8 MAGNESIUM

Magnesium has some essential functions in plants. It is involved in both physical and biological activities, including protein synthesis, carbon dioxide fixation, ATP generation in the chloroplast, and photooxidation in leaves. Mg deficiency in soils causes increasing concerns for high-productivity agriculture [22]. With a decrease in Mg, intervascular chlorosis and a decrease in growth are observed. Additionally, it affects the reduction of photosynthesis [23].

3.9 ZINC

Plant roots take up zinc as Zn^{2+}. ADH, an enzyme that catalyzes the conversion of acetaldehyde to ethanol, is one of the several enzymes that zinc activates. The pH of the soil, the existence of other macro- and micronutrients, and soil microbes can all affect the availability of enough quantity of zinc [5]. Zinc plays a vital role in helping plants gain resistance to various diseases, ensuring the integrity of the cell membrane, photosynthesis, protein synthesis, and increasing the level of chlorophyll. It is required at minimal but critical concentrations for several physiological functions of plants, such as sugar and photosynthesis, fertility and seed formation, growth regulation, and disease resistance [24]. Auxin hormone metabolism is directly impacted by zinc deficiency in the soil. A low amount of zinc causes the rate of photosynthesis to decrease by 50–70%. This decrease in photosynthesis may be proportional to the decrease in the carbonic anhydrase enzyme [24].

3.10 CONCLUSION AND FUTURE PERSPECTIVE

Sustainable agriculture is widely discussed in the world agenda, and studies are being carried out to minimize the impact of changes in the ecosystem. The continuation of sustainable agriculture can be achieved by reducing the germination time of the products, increasing the yield percentage, and protecting the plants from abiotic and biotic stresses. In this review, we discussed various nutrients necessary for the development of the peppermint plant. If the nutrients required for plant metabolism are not provided by farmers, plant development, product quality, and productivity decrease significantly. Research on various plant enzymes and plant transporters can fully reveal the roles of micro- and macronutrients.

KEYWORDS

- **peppermint**
- **mineral**
- **growth**
- **yield**
- **macronutrient**

REFERENCES

1. Parrey Z. A., Shah S. H., Fayaz M., Casini R., Elansary H. O., Mohammad F. Nitrogen supplementation modulates morphological, biochemical, yield and quality attributes of peppermint. *Plants*. 2023, 12(4), 809. https://doi.org/10.3390/plants12040809
2. Mamadalieva, N. Z., Hussain, H., Xiao, J. Recent advances in genus Mentha: Phytochemistry, antimicrobial effects, and food applications. *Food Front*, 2020, *1*, 435–458.
3. Tepecik, M., Esetlili, B. Ç., Öztürk, B., Anaç, D. Effect of different fertilizers on peppermint—Essential and non-essential nutrients, essential oils and yield. *Italian Journal of Agronomy*, 2022, *17*, 75–84.
4. Bouain, N., Krouk, G., Lacombe, B., Rouached, H. Getting to the root of plant mineral nutrition: Combinatorial nutrient stresses reveal emergent properties. *Trends in Plant Science*, 2019, *24*, 542–552.
5. Mondal, S., Bose, B. Impact of micronutrient seed priming on germination, growth, development, nutritional status and yield aspects of plants. *Journal of Plant Nutrition*, 2019, *42*, 2577–2599.
6. Khan, M. I. R., Nazir, F., Maheshwari, C., Chopra, P., Chhillar, H., Sreenivasulu, N. Mineral nutrients in plants under changing environments: A road to future food and nutrition security. *Plant Genome*, 2023, *16*(4), e20362.
7. Andresen, E., Peiter, E., Küpper, H. Trace metal metabolism in plants. *Journal of Experimental Botany*, 2018, *69*, 909–954.
8. Yang, X., Alidoust, D., Wang, C. Effects of iron oxide nanoparticles on the mineral composition and growth of soybean (*Glycine max* L.) plants. *Acta Physiologiae Plantarum*, 2020, *42*, 1–11.
9. Xu, G., Magen, H., Tarchitzky, J., Kafkafi, U. Advances in chloride nutrition of plants. *Advances in Agronomy*, 1999, *68*, 97–110.

10. Roriz, M., Carvalho, S. M. P., Vasconcelos, M. W. High relative air humidity influences mineral accumulation and growth in iron deficient soybean plants. *Frontiers in Plant Science*, 2014, *5*, 1–8.
11. Leghari, S. J., Wahocho, N. A., Laghari, G. M., Hafeez Laghari, A. Role of nitrogen for plant growth and development: A review. *Advances in Environmental Biology*, 2016, *10*, 209–218.
12. Wang, F., Wang, L., Yu, X., Gao, J., Ma, D., Guo, H., Zhao, H. Effect of planting density on the nutritional quality of grain in representative high-yielding maize varieties from different eras. *Agriculture*. 2023, *13*, 1835.
13. Iqbal, N., Umar, S., Khan, N. A. Nitrogen availability regulates proline and ethylene production and alleviates salinity stress in mustard (*Brassica juncea*). *Journal of Plant Physiology*, 2015, *178*, 84–91.
14. Lv, X., Ding, Y., Long, M., Liang, W., Gu, X., Liu, Y., Wen, X. Effect of foliar application of various nitrogen forms on starch accumulation and grain filling of wheat (*Triticum aestivum* L.) under drought stress. *Frontiers in Plant Science*, 2021, *12*, 1–17.
15. Aydin, M., Tombuloglu, G., Sakcali, M. S., Hakeem, K. R., Tombuloglu, H. Boron alleviates drought stress by enhancing gene expression and antioxidant enzyme activity. *Journal of Plant Nutrition and Soil Science*, 2019, *19*, 545–555.
16. L PA. Chapter 6. *Boron Deficiency*, 1990, 63–78.
17. Hasanuzzaman, M., Fujita, M., Oku, H., Nahar, K., Hawrylak-Nowak, B. Plant nutrients and abiotic stress tolerance. In: *Plant Nutr Abiotic Stress Toler*, 2018, pp. 1–590.
18. Razaq, M., Zhang, P., Shen, H. L., Salahuddin. Influence of nitrogen and phosphorous on the growth and root morphology of Acer mono. *PLoS One*, 2017, *12*, 1–13.
19. Wang, M., Zheng, Q., Shen, Q., Guo, S. The critical role of potassium in plant stress response. *International Journal of Molecular Sciences*, 2013, *14*, 7370–7390.
20. Sakhonwasee, S., Phingkasan, W. Erratum to: Effects of the foliar application of calcium on photosynthesis, reactive oxygen species production, and changes in water relations in tomato seedlings under heat stress. *Horticulture, Environment, and Biotechnology*, 2017, *58*, 310, https://doi.org/10.1007/s13580-017-0194-1.
21. White, P. J. The pathways of calcium movement to the xylem. *Journal of Experimental Botany*, 2001, *52*, 891–899.
22. Cakmak, I., Yazici, A. M. Magnesium: A forgotten element in crop production. *Better Crop*, 2010, *94*, 23–25.
23. Jin, X., Ackah, M., Wang, L., Amoako, F. K., Shi, Y., Essoh, L. G., Li, J., Zhang, Q., Li, H., Zhao, W. Magnesium nutrient application induces metabolomics and physiological responses in mulberry (Morus alba) Plants. *International Journal of Molecular Sciences*, 2023, *24*, 9650.
24. Rudani, K., Patel, V., Prajapati, K. The importance of zinc in plant growth-a review. *Journal of Natural and Applied Sciences ISSN*, 2018, *46*, 2349–4077.

CHAPTER 4

Abiotic Stress Alleviation in Peppermint Using Conventional Plant Growth Regulators

NEETIKA KIMTA,[1] RAJNI DHALARIA,[2] SHIVANI GULERIA,[3] SIVAKUMAR MANICKAM,[4] KAMIL KUČA,[5] RICHARD CIMLER,[6] and HARSH KUMAR[7,*]

[1]*School of Biological and Environmental Sciences, Shoolini University of Biotechnology and Management Sciences, Solan, Himachal Pradesh, India*

[2]*Department of Biosciences, Minerva P.G. College of Arts, Science and Commerce, Indora, Himachal Pradesh, India*

[3]*Department of Biotechnology, TIFAC-Centre of Relevance and Excellence in Agro and Industrial Biotechnology (CORE), Thapar Institute of Engineering and Technology, Patiala, India*

[4]*Chemical and Energy Engineering, Faculty of Engineering, Universiti Teknologi Brunei, Bandar Seri Begawan, BE1410, Brunei*

[5]*Centre for Basic and Applied Research, University of Hradec Kralove, Hradec Kralove, Czech Republic*

[6]*Centre of Advanced Technologies, Faculty of Science, University of Hradec Kralove, Rokitanskeho 62, Hradec Kralove, Czech Republic*

[7]*Centre for Advanced Innovation Technologies, VSB-Technical University of Ostrava, Ostrava-Poruba, Czech Republic*

*Corresponding author

Peppermint: Novel Strategies for Performance Improvement. Bilal Amhad, Tariq Ahmad Dar, and Khalid Rehman Hakeem (Eds.)

DOI: 10.1201/9781779641557-4

ABSTRACT

Abiotic stressors namely cold, salinity, severe temperatures, water scarcity, heavy metal, and metalloid toxicity usually limit plant growth in peppermint. The negative consequences of these stresses result in low chlorophyll concentration, which has a direct impact on photosystems, causing altered glucose metabolism and triggering secondary strains such as nutritional stress, and oxidative damage, resulting in decreased plant development. Plant growth regulators (PGRs) are signaling elements generated inside plants. Under climatic or stressful conditions, biosynthetic pathways normally regulate development and growth that either operate at synthesis site or are transferred to another plant parts. Several phytohormones, comprising abscisic acids, auxins, brassinosteroid, cytokinin, ethylene, and gibberellin regulate developmental activities and plant growth. However, new chemicals such as strigolactones, nitric oxide, and polyamines are now included in this list of PGRs in the past few years. PGRs are essential for expansion and progress of plant, and influence functional and biochemical processes of peppermint. Researchers have stated the importance of PGRs in reducing abiotic stress in plants. This chapter's main goal is to emphasize how various plant-growth regulators can be used to lessen abiotic stress in peppermint. Numerous signaling pathways are also highlighted, which are engaged in response mechanisms and improve plant survival in harsh environmental conditions.

4.1 INTRODUCTION

Mentha piperita L. (peppermint) is an herb of the family Lamiaceae and possesses medicinal and aromatic activities. Its leaves and oil are utilized as flavoring agents in traditional treatments and pharmaceutical products. Peppermint oil offers medicinal, antifungal, anticholinestearic, antibacterial, and antioxidant properties. Terpenoids, namely isomenthone, limonene, menthol, menthone, menthyl acetate, menthofuran, and pulegone, are the chief bioactive compounds [1–3]. The utmost damaging hazard to the environment, which is receiving significant consideration from academics, policymakers, and farmers, is global climate change, which has a significant impact on medicinal plants [4]. Drought, freezing, salinity, severe temperatures, flooding, heavy metal, and metalloid toxicity are all examples of numerous abiotic stressors that reduce productivity and impede plant growth [5].

Previous research has reported that salt stress has a deleterious impact on osmotic and ionic balance in plants, resulting affects on secondary metabolite development and production, such as components of essential oil [6–8]. The introduction of salinity-tolerant cultivars can ominously diminish the detrimental effects of salt stress, well-adjusted manuring, prudent fertilizer, and plant growth regulator (PGR) use [9, 10]. Researchers have proven that heavy metals are detrimental to the emergence and proliferation of plants, yet, in small amounts, a few metals are essential for plant development, and serve as critical components of many crucial chemicals [11]. The explicit reactions of plants under stress of heavy metal were a decrease in stomatal density and chlorophyll content [12]. A lowered chlorophyll concentration directly affects photosystem, causing disruption in glucose metabolism and triggering additional challenges, including stress caused by oxidation and stress related to nutrition, resulting in decreased vegetation expansion and output [13]. Increased concentration of salts in soil is a serious constraint to agriculture since it adversely affects plant development. Temperature is discovered to be the most important element influencing plant development rate. Pollination is considered as most phenologically vulnerable phase to increased temperature in plants, and assembly would suffer considerably throughout the developing period [14]. Extreme temperatures typically disrupt photosynthesis and bound carbohydrate supply, leading to kernel abortion that occurs during the reproductive period, when production of starch in the endosperm of grains is reduced due to a decrease in the action of many enzymes, producing dense grain [15].

The term PGRs, recognized as "phytohormones," are indicator fragments that are secreted by floras. PGRs are produced by the metabolic pathways in plants, typically facilitating progression and expansion routes that take place at the production location or are transferred to another area of the plant body in accordance with meteorological or adverse circumstances. Phytoregulators such as gibberellins (GAs), auxins (AUX), brassinosteroids (BRs), cytokinins (CKs), and ethylene (ETH) coordinate progress and expansion and control an array of biological and biochemical processes. However, more substances such as NO, polyamines (PAs), and SLs have also been included to this list in recent years [16]. GAs are natural diterpenes that play important roles in germination of seed, growth, bolting, flower induction, and the development of fruits [17]. Under abiotic stress circumstances, the hormone may trigger or enhance endogenous PGR levels [18]. An endogenous signaling molecule, salicylic acid (SA) regulates plant biochemical and physiological activities, such as photosynthetic and oxidative respiration [19–22]. It aids plant forbearance to salt stressors by inhibiting ETH biosynthesis, lipid peroxidation, modifying nutrient absorption, membrane functioning, and water relations [23–26]. Endogenous plant growth regulators govern responses of plants to abiotic stressors by sensitizing developmental and growth processes. The chief aim of the current chapter is to highlight the insights of PGRs in abiotic stress conditions. The function of ABA and CKs in stress tolerance, as well as their physiological and molecular mechanisms, is explained as well. The function of AUX, BRs, ETH, GAs, and PAs in stress tolerance mechanisms, moreover, their putative crosstalk routes in the development of stress tolerance, is gaining attention. New developments in hormonal signaling, mutant research, and physiological activities, as well as the compassion and characterization of the gene governing secretion of various endogenous progress controllers, point to the possibility of using their biosynthetic routes to create genetically engineered plants that are more resistant to environmental stressors [27].

4.2 ABSCISIC ACID

Abscisic acid (ABA) is also denoted as "the stress hormone" [28]. ABA is a phytohormone (isoprenoid) that is obligatory for the appropriate expansion and growth of plants, regulates physiological reactions ranging from protein storage to stomatal opening, and permitting reaction to a variety of inducements, including salt, drought, and cold [29]. The involvement of ABA in stress forbearance is widely documented and as a key transmitter in plants' modifiable reaction toward environmental stress has got immense attention. According to O'Brien and Benková [30], ABA innermost levels rise rapidly under environmental stress, influencing gene expression patterns and activating certain signaling pathways. Furthermore, the primary effect of ABA is to modify root architecture [31]. ABA-mediated responses will be impacted by changes to the root environment [32]. ABA is thought to be the master a plants capability to endure abiotic stress is monitored by harmonizing a number of processes [31]. Several genes from the ATP-binding cassette family that encode ABA vehicles have been identified in vascular tissues. AtABCG25 is a component of the genome that expresses an ABA messenger protein. It transports ABA from the interior of the cell to the outside. Guard cells have been discovered to carry another transporter, AtABCG40, which allows ABA to enter the cells from the outside. Both of these genes have been linked to ABA signaling. Numerous studies have been performed in an effort to further investigate the principles and essential aspects that contribute to ABA signaling under wide-ranging stress-related circumstances. The PP2C enzyme, an adverse inhibitor, and (sucrose nonfermenting) are all PYR/ PYR1-like monitoring components of the ABA receptors. SNF1-related protein kinase 2 (SnRK2) is an optimistic controller that is all engaged in ABA signaling. These three components combine to produce the "ABA signalosome," a signaling complex that functions as a two-tiered negative regulatory structure [33].

4.2.1 ROLE OF ABSCISIC ACID IN ABIOTIC STRESS SIGNALING

The administration of ABA to *M. piperita* simulates the effects of stress. Because numerous abiotic stimuli occur in osmotic imbalance and cell desiccation, the pattern of expression for stress genes in response to cold, dryness, excessive salt, or ABA administration overlaps. This implies that multiple stressors and ABA share signaling pathways, pathway elements, and several elements in common interact with one another to preserve cellular homeostasis. A wide range of stress cues causes an increase in ABA levels [34]. Once the conditions for the seed to germinate and flourish are perfect does ABA enables the seeds to overcome stress and germinate. Premature embryo germination is also prevented by ABA. Stomatal closure inhibits intracellular water loss during drought circumstances. Stomatal closure inhibits intracellular water loss in dry situations; hence, ABA is accurately regarded to be a stress hormone.

ABA appears to serve a primary purpose in terms of tolerance to osmotic stress and regulating plant water balance. If ABAs are extended as the drought lasts, ABA-deprived mutants wilt and perish. Under salt stress, ABA-deficient mutants grow poorly as well [35]. ABA is also essential to tolerate cold stress, which includes the activation of genes responsible for tolerating dehydration [35]. According to researchers, excessive salt or drought stress causes osmotic stress, which can be expressed by at least two processes, one ABA-associated and one ABA-unassociated. The possessions of cold on gene expression are mostly mediated by an ABA-independent mechanism [36, 37]. The persistence of an ABRE element (ABA-responsive element), a cis-activating element, is typically required for ABA-induced expression [36–38]. Calcium's participation in the *P5CS* gene is activated by ABA during salt stress [39]. Los5 mutants have been shown to have reduced countenance of the *COR15A, COR47, P5CS, RD22,* and *RD29A* genes. The signaling mechanism responsible for these genes' activation is unknown, although the *RD29A* transcriptional inducement of a few stress-associated genes is widely characterized [34]. The ABA-independent pathway was followed by calcineurin B-like protein (CBL) and CBL-interacting protein kinase (CIPK). The RD29A inheritable factor transcript accumulation testified to be managed in ABA-associated and ABA-unassociated ways [40]. Plants can accumulate proline *via* ABA-associated and ABA-unassociated signaling mechanisms [41]. The increase in mRNA for DNA helicase 45 (PDH45) generated by salt-induced damage followed an ABA-mediated pathway [41, 42]. Overall, the majority of ABA-dependent mechanisms mediate osmotic stressors' gene assertion. Lately, Lee et al. hypothesized that polymerized AtBG1 stimulation of latent ABA pools could be a technique through which *M. piperita* rapidly adjusts ABA levels in response to environmental change stimuli [43].

4.3 AUXINS (AUX)

AUX is also well known among phytohormones for influencing all elements of plant root enlargement under favorable and unfavorable conditions [44]. Root development, apical dominance, the formation of auxiliary buds, the development of floral organs, and the differentiation of vascular bundles are all controlled by AUX [45–47]. AUX (indole-3-acetic acid, IAA) regulates apical dominance, tropic reactions, elongation of cells, division of cells, proliferation, and sprouting of roots in plants, among many other activities. Recent research has proven that AUX is indispensable for plants' responses to unfavorable abiotic and biotic situations. The assertion of genes convoluted in AUX-related routes varies depending on the environment. Indeed, substantial acumens into the crucial character of AUX under osmotic stress triggered by scarcity of water, salinity, and extreme temperature have come from the functional activities of AUX-related genomes in plants that have been revealed to be vulnerable to abiotic stresses [48]. Two putative pathways for stress-induced changes in IAA levels in *M. piperita* have been proposed: one involves changes in the countenance of the AUX polar vehicular gene, and the other involves the deactivation of polar transport by certain chemicals accumulating in reaction to stress [27]. The countenance use of the AUX-forthcoming marker IAA2-GUS in conjunction with the undeviating AUX vehicular test in the context of chilling stress-prompted decreases in root development and magnitude response revealed that cold stress largely directed intracellular AUX conveyance. Cold stress has a close relationship to the intracellular AUX incline, which is mediated by intracellular AUX vehicle operating and polar dispersion. Some evidence of AUX transport carriers' involvement under drought stress conditions has been found. PIN proteins were assumed to perform a direct or indirect function in drought stress as a critical regulator of AUX transport in plants (Figure 4.1) [49].

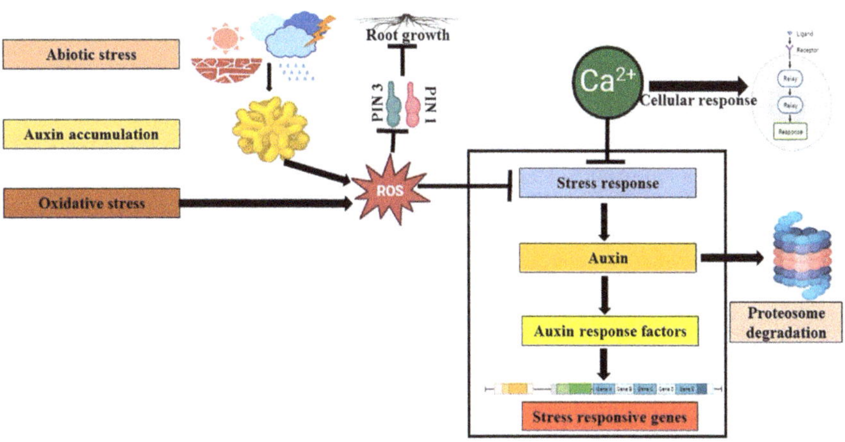

FIGURE 4.1 Auxin signaling mechanism under abiotic stress conditions [48] (License no. 5625440007662).

4.3.1 AUXIN SIGNALING UNDER ABIOTIC STRESS

Abiotic stressors cause many phytohormones, including AUX, to vary in endogenous levels. Plants must maintain an appropriate level of endogenous AUX to coordinate their many cellular functions (Figure 4.1). In plants, AUX can be found both unrestricted and restricted (to various amino acids, carbohydrates, and peptides) [50]. The processes that control AUX homeostasis in plants are biosynthesis, degradation, and conjugation [51]. The most crucial methods for maintaining the endogenous AUX pool are AUX production and distribution in conjugated form to various plant organs [52, 53]. IAA conjugates aid in IAA transportation and act as a short-term supply of inactive IAA that can be digested to produce active hormone for plants. Changes in IAA levels and relocation in response to environmental cues affect plant progression and expansion. Under abiotic stressors, variations in AUX transport may control local AUX concentration and AUX dispersion [54]. In retort to cold stress, OsGH3-2, an adherent of the GH3 family that catalyzes the conjugation of IAA to amino acids, displayed elevated expression [55].

Plants that overexpressed OsGH3-2 exhibited morphological abnormalities linked to IAA shortage and had lower levels of ABA, carotene, and free IAA, which improved their ability to withstand cold temperatures. According to Du et al. [55], this impact was caused through the regulation of endogenous free ABA homeostasis and IAA levels. Low AUX mutant lines that lack carotenoids showed resistance to cold stress [56]. The tryptophan-dependent AUX biosynthesis pathway's flavin monooxygenase gene, encoded by the *YUCCA6* gene, overexpressed plants displayed high AUX characteristics, including narrow kinky leaves, greater elevation, tall stature, and permanency [57]. Similarly, activation of the *YUCCA7* gene, which encodes the flavin monooxygenase, primes to an increase in total AUX levels and shows plant characteristics resembling those seen in AUX overproducing mutants [43]. These findings imply a connection between plant responses to abiotic stress and endogenous AUX levels.

4.4 BRASSINOSTEROIDS

Brassinosteroids (BRs) are newly discovered poly-hydroxylated nonintrusive, versatile steroidal plant hormones, and novel [58]. The efficient BRs are generally used in plant biochemical and physiochemical research. These compounds comprise brassinolide (BL), 24-epibrassinolide (24-EBL), and 28-homobrassinolide (28-HBL) [59]. According to Bhandari and Nailwal [58], BRs encourage floral reproduction, fruit maturing, leaf ageing, rhizome formation, maturation of roots, sprouting of seeds, vegetative growth, and stomatal development are all processes that occur in plants, all examples of plant processes. Different biological and structural processes in *M. piperita* are positively mediated by BRs. According to Karlidag et al. [60], adding BRs under salt stress greatly improved their growth characteristics, antioxidant enzymes, chlorophyll concentration, transmembrane integrity, and elemental content are all factors to consider. The most prominent BR-mediated reactions in plant responses were increased CAT, gas exchange elements, and osmolyte content (SOD, POD, and pigment). The findings imply that BRs are potentially beneficial, environmentally benign, naturally befalling elements that are employed extensively to mitigate the possessions of abiotic stress.

4.4.1 MODE OF BRASSINOSTEROIDS ACTION UNDER ABIOTIC STRESS

Although it has been a desirable target for many researchers, the mechanism of action of BRs is still far from being resolved. There are two key elements of the fundamental mechanism to take into account. First is how BRs affect gene expressions, which in turn affects how different enzymes are synthesized. Second is how BRs affect the characteristics of membranes. It is generally known that steroid hormones serve as signaling agents in both plants and animals. In plants, BR-insensitive 1 (BRI1) is a kinase of cell surface receptors that recognizes BRs. Between the two most efficient BRs in bioassays, namely EBL and HBL, interact to the intracellular region of the BRI1 receptor. According to Friedrichsen et al. [61], a serine/threonine (S/T) kinasess BRI1 receptor is located on the plasma membrane. An isolated domain is a length of amino acids that connects LRRs 21 and 22, which is another distinctive characteristic of BRI1 that is necessary for BR binding [62]. The minimal binding site for BRs is made up of this domain and the neighboring LRR22. When BRI1 is bound by BL, it communicates with the proteins associated with BRI1 receptor kinase 1. Along with its intracellular domain, BRI1 is phosphorylated many times, some of which have been found to control receptor activation [63]. The BR-insensitive 2 (BIN2) is a detrimental controller that modulates the BR biogenesis route, and is inhibited by the BR signal after it is conveyed to the cytoplasm by an unidentified method. The brassinazole-resistant 1 (BZR1) and gene transcription regulators BRI 1-EMS suppressor 1 (BES1) are nearly identical and are negatively controlled by the protein kinase BIN2 through interactions and phosphorylation. BES1 and BZR1 are dephosphorylated by BRI 1 suppressor 1 to mitigate the adverse effects of BIN2. BRs standardize the countenance of several thousand of genes. It is anticipated that a sizable number of the uncontrolled genes will be involved in growth activities. BES1-interacting MYC-like1 (BIM1) works in coordination with BES1 to increase both the binding activity of BES1 and the expression of its target genes. BIM1, BES1 dimerize with another transcription factor, increasing its activity. Several BR biosynthesis genes are among the genes that BRs downregulate. BZR1 interacts target specific elements associated with the promoters to reduce their activity. The genes that BZR1 suppress show an unfavorable feedback mechanism for BR growth regulation [64, 65].

4.5 CYTOKININS

Cytokinins (CKs) are growth-promoting phytohormones that are found in nature and regarded as an organogenesis master regulator [66]. According to Cortleven et al. [67], CKs control fruit and leaf abscission as well as biogenesis, cell division, chloroplast morphogenesis, root formation, seed germination, and shoot differentiation are all processes that occur in plants. When a plant is under stress, other hormones, such as CKs, which manage a variety of critical plant growth processes, are also released. According to research by O'Brien and Benková [30] and Zwack and Rashotte [68], endogenous CK levels are found to change in reaction to a multiplicity of circumstances, comprising xeric and excessive salinity. The manufacturing and transportation of CKs from roots are hampered by these harsh climatic circumstances. Exogenously administered CKs can increase stomatal apertures, transpiration, and positively impact photosynthetic activity in plants, demonstrating that they may join in controlling how plants adapt to ecological challenges. When assessing the role of CKs in stress reactions, a number of physiological activities must be considered, among them: cell proliferation, which is a highly energy-intensive process; photosynthesis's efficacy as an essential energy supplier; and activation of the sink efficiency [69]. As ABA suppresses seed germination, CKs are frequently thought of as ABA antagonists because they cause seeds to emerge from dormancy. Low concentrations of CKs encourage apical dominance, which limits stomatal aperture opening and aids plants in coping with drought stress, combined with an increase in ABA [70]. CKs are regarded as an imperative signal class and typically move from roots to shoots. Numerous plants have reported their xylem sap's CK content, which provides information about how that sap changes when a plant is experiencing drought. Under elevated CK levels, the primary gene for CK production, IPT, is regulated by the senescence-triggered SAG12 and is associated with appropriate inundation tolerance, and is highly expressed. SARK (senescence-associated receptor kinase) was revealed to impact IPT countenance in tobacco plants, consequential in extraordinary tolerance to severely dry conditions. SARK is a drought, maturation-induced promoter. These results and the role that CKs have been reported to play in sink-source polarization suggest that they play a part in how plants react to drought [49].

4.6 ETHYLENE

According to Alhaithloul et al. [71], ETH is a minuscule olefinic gaseous signaling molecule with a variety of functions. ETH, also known as the "aging" hormone, is distinctive among hormones because it is a gaseous hydrocarbon. Although it has importance in both the initial and late stages of plant expansion, ETH also controls growth over the plant's whole life cycle. It is linked to cell division and has the ability to control cell size, frequently preventing cell elongation. Along with these roles, ETH also promotes in many plant species, the rooting process begins, encourages flowering [72]. Under many temperature extremes, excessive rainfall, extreme heat, extreme ultraviolet (UV) radiation, high levels of salt, heavy metal, and organic pollutions are examples of environmental stresses; the biosynthesis of ETH rises. This phenomenon is regarded as "ETH stress." It has been discovered that a variety of stresses can start the amalgamation of ETH, which can then moreover amplify the signs of stress or start a signal cascade that leads to reactions that improve plant survival under challenging circumstances. Following the initiation of stress, this first ETH synthesis starts a defensive response that includes the transcription of genes engaged in disease and the resistance by acquisition. If stress condition continues, the ACC biogenesis will be activated (1-aminocyclopropane-1-carboxylic acid) gene will be transliterated. As a result of the additional ACC, ETH synthesis will increase once more, which will have negative impacts on plants and cause processes such as abscission, chlorosis, and senescence that could significantly reduce plant survival [73]. The primary regulator of plant development and senescence is the signaling pathways of ETH. According to Kalay et al. [74], ETH treatment dramatically increased shoot and root growth, as well as reduced cellular damage. As a result, ETH treatment is crucial in helping plants cope with a variety of abiotic stresses.

4.6.1 ROLE OF ETHYLENE UNDER ABIOTIC STRESS

Wounding is the abiotic stressors-ETH reactions that have been investigated the most. ETH levels rise in plants after mechanical or herbivore injury, and the ETH biosynthetic gene expression is changed [75]. ETH also manages the PDF1.2 gene's expression, which is JA-inducible, pathogen- and wound-responsive [76]. ETH emission rises under drought stress [77, 78]. Plant hormone ABA modulates the closure of stomatal cells and momentous role in drought-induced stress in plants. ACC insertion of the ETH transgenic mutant eto1 was used to demonstrate that ETH blocks ABA-induced stomatal closure, causing less stomatal closure after ABA treatment [79]. It is unclear how ETH affects osmotic stress. The phrase of the ETH receptor ETR1 is suppressed by osmotic stress, and this change is replicated at the protein level [80]. Under hypoxia, *Arabidopsis* plants have raised ACS2, ACS6, ACS7, and ACS9 mRNA levels. AOA (aminooxyacetic acid), a blocker of ETH production, impairs the control of ACS9, and ETH signaling mutants etr1-1 and ein2-1 possess reduced regulation of ACS9 [81]. The fact that ACS2 mRNA levels are decreased by AOA treatment but that ACS6 and ACS7 levels are unchanged, suggests that the regulation of ETH emission during hypoxia may be subject to complicated control [81]. Additionally, ethanol is noteworthy in how the body reacts to abiotic stimuli. A burst of ETH is produced when high levels of ozone are applied to plants [82], which promotes ozone-induced cell death [83]. This mechanism could involve SA production and accumulation [84] and the inhibition of JA's cell-protective action [85]. The damage brought on by ozone treatment is reduced when ACS activity in plants is suppressed [86, 87].

4.7 GIBBERELLINS (GAS)

Gibberellins (GAs), a growth hormone, are widely acknowledged for their functions in the growth of flowers and fruits, the lengthening of stems, the leaves commencement, and seed germination. GAs are also recognized to be noteworthy in managing numerous plant growth processes are caused by the impact that they have on photosynthetic enzymes, leaf entrapment, and the area of the leaf index, boosting photosynthetic efficiency and nutrient utilization [88]. After many years of research, we now have important information about how GAs function under stressful circumstances. Growth is restricted in several plant sections under various abiotic stresses *via* the main GA negative regulators, the DELLA proteins [89]. Under reduction in physiological amounts of bioactive GAs under salt stress has been noted, which is accompanied by a rise in DELLA protein accumulation. Salt stress impeded development and growth in one experiment, wild-type plants were used, whereas a DELLA quadrupled malfunctioning mutant was impaired by salinity stress in the context of reduced primary root growth, delayed blooming time, and slowed leaf formation. SLEEPY1 and SNEEZY (SNZ) are F-box proteins, as well as the GA receptor GID1 are important regulators of GAs signaling during stress responses. GA forms a complex with the DELLA proteins when it binds to the GID1 receptor, which inhibits nuclear development. SCFSLY1/GID2/SNZ then polyubiquines and degrades the DELLA protein. As a result, GAs encourage plant development by destroying DELLA proteins, whereas DELLA proteins work to limit plant growth [90].

4.7.1 GIBBERELLINS RESPONSES TO ABIOTIC STRESS

The discovery that DELLA proteins at least partially mediate growth restraint in response to diverse forms of abiotic stress [91–93] led to a substantial improvement in our comprehension of how GA controls plant development in response to stress. The ability to survive in salt stress has been connected to DELLA function [91, 92, 94]. GA scarce biosynthetic transformed ga1-3, for example, demonstrated increased survival under extreme salinity, through a quadruple-DELLA mutant being vulnerable [91]. It is significant to note that growth metrics and salt forbearance in an assortment of DELLA mutants revealed a substantial association amongst plant stature, duration to flowering onset, and vulnerability to salinity, showing that DELLA proteins might restrict progress as well as enhance stress forbearance *via* a single mechanism [92]. The effects of cold stress on seedlings are similar to those of salt stress in that they cause a decrease in activated GA, enhance DELLA production, and result in DELLA-triggered growth constraint [95].

Again, the DELLA function supported stress tolerance, as evidenced by the lower survival of freezing in DELLA mutants. ROS, which have been shown to undergo accretion in salt-treated plantlets and likewise in a diversity of extraneous abiotic and biotic challenges, were found to be inhibited by DELLA activity in both cases. This was attributable to transcription aspects from the C-repeat binding factor (CBF)/dehydration-responsive element binding protein (DREB1) family upregulating particular *GA2ox* genes. When ROS levels are high, plant cells die, and DELLA delays this reaction [92]. For Sub1A lines wide-open to dehydration after floods and to drought, a parallel link was discovered between sustained intensities of SLR1 and lower intensities of oxidative stress [96]. It was proposed that ROS might connect effects on development and resilience to stress mediated by DELLA because they appear to regulate GA-mediated root growth as well [92]. DELLA buildup is associated with growth constraint upon revelation to abiotic stress [89, 97], whereas under unstressed conditions, the amount of DELLA-mediated growth limitation appears to be related to stress survival [92]. DELLA is known to control cell proliferation [95, 98], and the buildup of DELLA causes mitotic exit when cells are exposed to mild osmotic stress [89]. However, under those circumstances, *M. piperita* with lower DELLA activity also showed a decrease in cell quantity (albeit the early beginning of endo-reduplication was prevented). This may point to a contrast between DELLA's role in stressful and nonstressful circumstances. Furthermore, evidence indicates revealed DELLA targeted different cell cycle controllers than was previously suggested under osmotic stress [89, 91]. A contrast between DELLA's role in stressful and nonstressful circumstances. Furthermore, evidence indicates revealed, in contrast to previously described cell cycle regulators DELLA targeted different regulators of the cell cycle amid osmotic stress.

4.8 JASMONATES (JA)

Methyl jasmonate (MeJA) and jasmonic acid (JA) are two examples of the multifunctional chemicals referred to as jasmonates (JAs) [99]. According to Dar et al. [100], JA controls several processes, including chlorophyll degradation, fruit ripening, germination of seed, leaf movement, leaf senescence, seedling progression, and root expansion. Plants need JA to endure stress, as is widely known, and adapt to it. Through ROS-mediated preservation of oxidative molecules, endogenous JA increased tomato salt tolerance [101]. Proline, soluble sugars, and growth characteristics were all greatly improved by JA treatment, while H_2O_2 concentration was decreased [102]. Whereas temperature variations affected the level of endogenous SA. In a different instance, Bali et al. [103] found that adding JA under lead poisoning increased growth features, ascorbate glutathione pathways, gas exchange components, metal chelating compounds, osmolytes content, and decreased H_2O_2 level while decreasing Pb uptake. When plants become vulnerable to abiotic stressors, challenges such as cold [104], salinity, drought, and heavy metal toxicity [105], JA activates plant defense systems. According to Dar et al. [100], they can lessen a variety of environmental hazards. In addition, JA regulated antioxidant system function *via* dropping the expression of P-type ATPase and RBO vehicular genes [103]. A lower freezing injury index resulted from MeJA application because it reduced phenolic levels and increased antioxidant enzyme activity [106]. Therefore, it is conceivable to draw the conclusion that JAs play a favorable monitoring character in abiotic stress response in plants.

4.9 NITRIC OXIDE

An important gasotransmitter called NO serves as a messenger element for plants under stress. NO intercellular transmission of stress signals occurs via crosstalk with other signaling molecules. This shows that crosstalk, often known as the sharing of a similar pathway by two or more biosynthetic pathways to better regulate signals, may exist. The aforementioned has been documented that there are several endogenous progression modulators, namely ABA and GA, which have the ability to stimulate seed germination and seed dormancy breakage. Phytohormonal crosstalk has been shown to be a mechanism by which azide, hydroxylamine, nitrate, nitrite, NO, and sodium nitroprusside chemicals influence seed dormancy and germination processes in recent years [107]. According to Desikan et al. [108], NO controls the closure of stomata through a $Ca^{[2+]}$-dependent mechanism. Research has also demonstrated the occurrence of NO in guard cells, which causes stomata to close as a result of NR activity [109]. Recently, it has been proposed that UVR8, H_2O_2, and NO interact with one another under UV radiation and control the UVR8 pathway, which eventually leads to stomatal closure [110]. Under biotic and abiotic stressors, rapid NO production and simultaneous accumulation of ROS are frequently seen. As a result, these unfavorable reactions start the senescence process, which ultimately results in plant cell death. According to earlier research, NO and ROS may work together or separately to regulate programmed cell death (PCD) [111]. As a result, NO serves important roles as a signaling chemical in the equilibrium of nutrients, ion transport, plastid development, and antioxidant gene alleviation in both favorable and adverse situations. Due to unwanted anthropogenic activities in the environment, abiotic stress is steadily rising [112]. By generating these stressors, known as ROS (hydroxyl radicals, hydrogen peroxide, single oxygen, superoxide radicals, and so on), have an unfavorable consequence on the expansion and development of plants [113, 114]. NO is essential for intracellular redox signaling, ion homeostasis, and antioxidant defense system activation are all involved. It has been classified as either a stress-inducing or protective mediator [112]. Through crosstalk with other signaling molecules, several studies have suggested that NOs contribute to the maintenance of pigment configuration, stomatal actions, root maturation and growth, water interrelationships, the integrity of the membrane, circadian rhythms, hydrological amendments, and ion channel actions in plants were studied under various conditions [115–117].

4.9.1 NITRIC OXIDE SIGNALING UNDER ABIOTIC STRESS

The plant life cycle depends on the ability to recognize stimuli coming from both inside and outside the cell. Plant signaling enables this perception. Through the use of signaling molecules, information is transferred between plant cells *via* plant signaling from receptors to effectors. Evidence for signal response as a result of crosstalk among numerous routes of signaling has been made possible by the discovery of molecular components relevant to signaling [118]. This crosstalk typically occurs from the combination of the particular signal response and pathway integration. Such physiological crosstalk included development as well as stress reactions [119]. ROS, PGRs, and the signaling peptides, which use NO as its primary signaling molecule, are a few of the essential signaling substances. NO manages numerous signaling reactions in plants. These reactions control the progression and expansion of plants at the intra and intercellular levels. These procedures are controlled by NO-mediated changes in transcription or secondary mediator activation [120]. Seed germination, cell wall lignification, hypersensitive reaction, PCD, organelle motility, blooming and pollen tube growth, fruit maturing, and biotic and abiotic stress are examples of these [121, 122]. NO signaling influences a variety of processes, including seed dormancy, plant reproduction mechanisms, and plant–pathogen interactions [123]. It also works on multiple levels, particularly with ROS from the antioxidant system [123]. Additionally, a weakened antioxidant system in plants is correlated with greater NO/ROS levels [124]. For nutrition and plant immunology, this NO/ROS homeostasis interaction is also essential. These activities are primarily controlled by NR activity, which, after stress induction, is a crucial component of NO signaling.

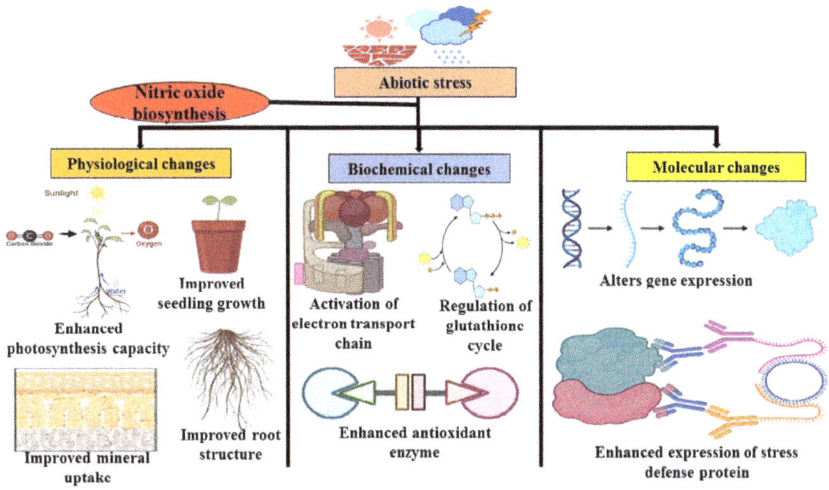

FIGURE 4.2 Biochemical, molecular, and physiological changes are balanced by nitric oxide under abiotic stress conditions.

Source: Reprinted with permission from Ref. [125]. Copyright © 2015 Springer Science+Business Media New York

According to Sivakumaran et al. [126], hormonal modulation of NO/ROS balance is an important module of plant development and stress reaction. By adjusting hypoxic or anoxic circumstances, mitochondria also serve a vital role in NO and ROS signaling [127]. In addition to NO's direct ROS creation role, posttranslational modification of NO enzymes, particularly the ascorbate-glutathione succession, is crucial for maintaining the balance between NO and ROS [128]. ROS levels are also impacted by NO self-regulation [129]. This regulation shows that NO and ROS are being adjusted as signaling elements. Another aspect of NO signaling is that it functions in the presence of additional signaling components such as cytosolic Ca^{2+} levels, phosphatidic acid, JA, cyclic guanosine 5′-monophosphate (cGMP), H_2O_2, cyclic adenosine diphosphate ribose, mitogen-associated protein kinases, and SA, it acts as a secondary messenger [130, 131]. Plants' NO-cGMP-dependent system creates opportunities for NO and cGMP signaling to interact [132]. Although NO-facilitated cGMP signaling is well understood in other beings, it is little understood in plants. The molecular basis for physiological and developmental responses brought on by NO signaling is provided by this interaction. Additionally, investigations on downstream target proteins provide hints about the NO signaling's indirect influence [133]. Aside from the traditional pathway, NO interacts with other compounds in the plant directly and affects biological processes (Figure 4.2). Examples include NO interactions with sulfur [134], inositol [45], heme oxidase [135], and H_2O_2. Its mechanism will be better understood if we consider that this interplay occurs in the circumstances of NO response and signaling.

4.10 POLYAMINES (PAS)

The main polyamines (PAs) discovered in low molecular weight plants include putrescine, spermidine, and spermine. According to Groppa and Benavides [136] and Chen et al. [137], they control a wide range of various leaf degeneration, cell proliferation and cell differentiation, fruit ripening and development, flower and gene expression, zygote formation, morphogenesis, and organogenesis are all physiological processes. Environmental factors such as water shortages, severe temperature, cold, and heavy metal stress all have a substantial impact on PAs [138]. Additionally, PAs effectively increased pigment, phytochelatin, and soluble sugar content while lowering H_2O_2 and MDA levels in chromium-toxic environments [139]. Under water stress, Hassan et al. [140] found that treatment with spermidine and spermine significantly increased growth attributes, gas alteration components, proline content, SOD, and CAT enzyme bustle. Exogenous spermidine and spermine supplementation boosted spermidine and spermine levels while decreasing endogenous putrescine levels under water stress conditions. Recent studies have thoroughly investigated the mode of response of PAs to the use of transgenics, PA production inhibitors, and administered PAs approaches, which are crucial for plant development. According to prior research, PAs have been linked to plant survival, stress resistance, nucleic acid and membrane structure integrity, and plant development [137]. Numerous exogenous PA supplementation, to varying degrees, has been shown to sustain the structural integrity of plant cell membranes, reverse or decrease stress-induced growth inhibition, and modulate the expression of osmotically responsive genes, decrease H_2O_2 and O_2 level, and increase antioxidant enzyme activity.

4.11 STRIGOLACTONES

Strigolactones (SLs) are a type of plant hormone represented by derivatives of carotenoids, are recently discovered as multifunctional compounds [141]. They have a root architecture, the root zone framework, and seed germination are all important in plants [142]. In addition, they control nutrient intake, stomatal closure, shoot branching, stem thickening, and leaf senescence [143]. Under drought stress, by considerably lowering oxidative damage, stomatal opening, and EL level, foliar application of SLs drastically decreased growth, levels of chlorophyll, fluorescent chlorophyll components, RWC, gas flow parameters, and protective enzyme activity. Furthermore, the utilization of SLs encourages genomic transcription essential for the fabrication of SLs [144]. Additionally, exogenous SL promoted proline accretion, maintained mineral nutrient absorption, and osmotic balance, while increasing SOD and POD activities were increased, and oxidative stress caused by KCL toxicity was reduced. In addition, SLs, which act as endogenous growth regulators, improved the tolerance to abiotic stress [144, 145]. According to Visentin et al. [146], SL supplementation dramatically increased stomata sensitivity under drought stress. Plants are able to survive harsh weather circumstances thanks to the modulation of numerous hormone response networks by SL, which also lessens the injurious impression on crop output as a whole [147]. Overall, SL appears to have a considerable impression on the physiology of abiotic stressors in horticulture plants in reaction to various ecological stressors.

4.12 CROSSTALK AMONG PLANT GROWTH REGULATORS UNDER ABIOTIC STRESS

The sensing detection of environmental stress by several receptor molecules found in organelle cells and the cell membrane is the initial step in activating the signaling pathway [148]. Crosstalk between phytohormone and signaling molecules is defined as the point at within the cell, hormone and signal transduction cascades interact. This could be a regulator both upstream and downstream of the phosphorylation cascade that is activated by hormones and secondary messengers [149]. Although these signaling and transduction cascades directly influence other hormones' production in the cell, they also regulate gene expression [150]. Plant growth regulators that resemble phytohormones also interact with other PGRs to significantly alter the phenology of horticulture plants. According to Niharika et al. [151], the development stage of the plant lifespan relies on the harmonization and modulation of the biochemical principles of plant growth through extensive connectivity and cooperation of several hormone regulators under varied ecological circumstances. ABA is crucial in controlling stress levels in reaction to drought stress [149]. Both the biosynthesis and gene expression of ABA may be regulated. Both the ABA 8-hydroxylation (CYP707A) JA has been shown to affect the genes for and 9-*cis*-epoxycarotenoid dioxygenase (NCED), which has a favorable effect in stressful situations [152]. Stomatal closure is a foremost morphological characteristic that is tightly modulated and is predominantly influenced by water scarcity. In drought-stressed situations, ABA may work along with NO and JA to encourage stomatal closure. In addition, drought stress affects the pattern of expression of genes associated with CK, ETH, and AUX. Crosstalk between IbARF5 and AUX and ABA increased proline synthesis and SOD activity, making it more resilient to numerous stressors like salt and drought conditions [153]. Previous studies have demonstrated that dopamine and catecholamines modulated and interacted with conventional hormones like ETH and AUX to control the metabolism of sugars and carbohydrates, which is indispensable for plant progression and expansion [154]. Dopamine was shown to stimulate the synthesis of ETH and decrease the oxidation of IAA in the apple plant when exposed to salinity stress [155]. Additionally, it was shown that dopamine and GAs work synergistically, which improves the plant's GAs-mediated reaction. Dopamine has also been found to be a powerful water-soluble antioxidant that may scavenge diphenylpicrylhydrazyl radicals and prevent linoleic acid from absorbing oxygen [156]. ABA is a well-known critical regulator of a plant's capability to withstand drought. The enzyme that limits the pace of the production of ABA is called NCED [157]. According to recent research, *Vitis vinifera* may tolerate drought when the *VaNCED1* gene from *Vitis amurensis* is overexpressed [158]. Additionally, the VaNCED1 overexpression lines increased the amount of JA by increasing genes convoluted in JA production, such as 12-oxophytodienoate reductase and allene oxide cyclase [158].

DELLA controls the GA signaling pathway negatively. It has been demonstrated that the transcription elements BZR1 and BR-INTENSIVE 1 EMS-SUPPRESSOR1 interact with DELLA to control the synthesis of GA, which is also controlled by BR. The BR signaling mechanism is likewise positively regulated by both transcription factors [151]. The enzyme GA2ox is a key speed-regulating enzyme in the GA biosynthesis pathway and a vital regulator of increasing plant development [160]. According to Shi et al. [161], StGA2ox1 overexpression in potatoes increased the transgenic potato plants express WRKY1, RD28, DREB1, and SnRK2. These findings reveal that GA interacts with transcription aspects that are controlled by the ABA transmission cascade. As a result, StGA2ox1 overexpression opens the door to offering opposition to circumstances, including water scarcity, salinity, and low-temperature stress [161]. Under abiotic stress, ETH interaction with other hormones is also essential. According to reports, melatonin and dopamine regulate nitrogen metabolism and absorption at low ammonium levels [162]. This could be because dopamine and melatonin enhanced the levels of GS, NR, and NIR enzymes in foliage and plant roots subjected to decreased NH_4^+ deliberations. GOGAT is present in higher plants in two forms: NADH-GOGAT takes up NH_4^+, whereas Fd-GOGAT is found in oxygen-deprived organs, takes up NH_4^+ in leaves. According to Per et al. [163], StEREBP1 overexpression improved plant lenience to both cold and salt stress. The EIN3/EIN3-like proteins are another descending member of the ETH signaling pathway that is regulated by abiotic stress in plants [164]. The ETH biosynthesis pathway was also controlled by MnEIL3, which also increased saline stress and drought stress tolerance-promoting MnACO1 and MnACS1 promoter activity [165]. In contrast to cells cultivated in basic media, dopamine boosted cell proliferation in an IAA and kinetin-enriched media [166]. Exogenous dopamine has also been shown to stimulate progress and intensification of photosynthetic efficiency, enhance nutrient uptake, and decrease metabolic stress in plants exposed to abiotic stresses, namely salt, water scarcity, and nutrient deficit, in numerous studies [167]. By interrelating with DELLA and BZR1 proteins, transcription factors, including phytochrome interacting factor4, which regulate light-mediated physiological processes, have been found critically significant in GA and BR signaling [168]. Phytohormones like JA and CK interact negatively with one another. MeJA affects the quantity of CK oxidase gene activity and expression in the plant [163]. This observation demonstrates how the CK and JA interact. Additionally, under stressful situations, JA antagonistically regulates the CK, in turn, regulates the behavior of genes associated with leaf senescence. On a related note, it has been suggested that CK and ABA have a synergistic impact. In addition to seed sprouting, pollen production, fertilization, embryogenesis, maturation, ripening, abscission, and senescence, Pas are engaged in a range of tissue coagulation, organ formation, blooming, and vegetative development activities [169]. In contrast to spermidine, putrescine enhanced the countenance of genes that govern ABA fabrication while lowering the regulation of genes required for the synthesis of ETH, JA, and Gas. Spermine, whereas, inhibited the gene countenance of genes intricate in the biosynthesis of Gas and ABA while increasing the transcription of genes involved in the creation of ETH and JA [170]. In terms of hormone signaling pathways, spermine activity was linked to genes involved in AUX and CK signaling, whereas SA signaling was positively controlled by spermidine. Putrescine affected jasmonate signals in a way that ranged from neutral to favorable. According to Anwar et al. [171], Pas do not seem to have any impact on the signaling or synthesis of BRs. Due to their amphipathic makeup, serotonin and melatonin can pass through membranes. Despite being present in most plant organs, the molecules are more noticeable under stressful conditions. Numerous mechanisms controlled by growth regulators such as hormones, ROS, and others have been linked to melatonin. Crosstalk between serotonin and melatonin is predominantly mediated through molecules implicated in downstream signaling, including IAA, GA, CK, ABA, ETH, JA, SA, and NO. Additional research is needed to better understand how serotonin and AUX interact in response to varied environmental inputs. In light of melatonin"s function in the interaction with other biomolecules, management of its fabrication and turnover seems to be indispensable [172]. By altering SA, the translation, transcription factors, and co-regulators are active and located. SA-triggered redox changes, interceded by thioredoxins and glutaredoxins, modify the locality and DNA-requisite action of transcriptional modulators involved in JA-dependent gene suppression, namely the NPR1 and TGA transcription factors. In addition, SA can shift the location of delaying JA-receptive transcription features away from target genes in the cytoplasm or in nuclear compounds with suppressor proteins. The JA signaling pathway induces the deterioration of genetic factors that serve as activators. SA also influences JA-induced transcription [173]. The most significant finding is that DELLAs contribute to disease resistance can be achieved by enhancing JA detection and signaling are increased while SA generation and signaling are decreased [174]. The impact of DELLA proteins on SA/JA signaling through progress and expansion must also be investigated. Additionally, ROS regulation is influenced by DELLAs [175]. DELLAs are thought to impede SA signaling ROS levels fall because ROS and SA signaling are closely related in an autoamplification loop (Figure 4.3). According to

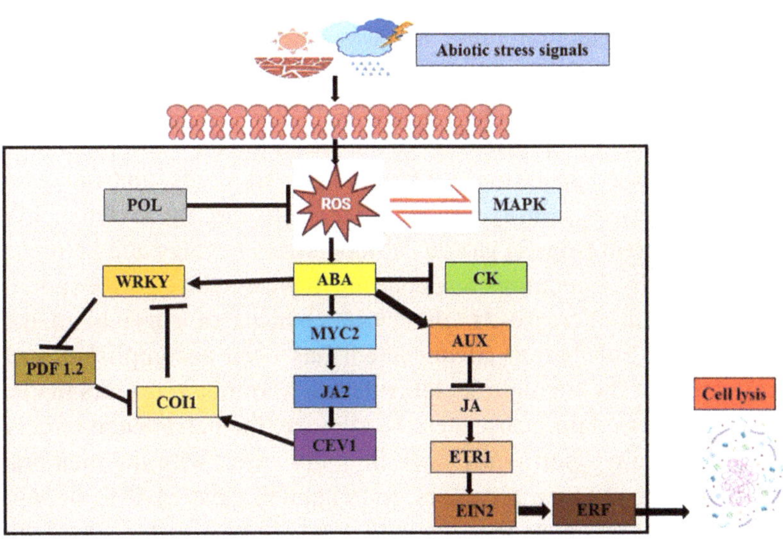

FIGURE 4.3 Crosstalk of signaling molecules and plant growth regulators exposed to abiotic stress.

Source: Reprinted with permission from Ref. [159]. Copyright © 2021, The Author(s), under exclusive licence to Springer Science Business Media, LLC, part of Springer Nature

crosstalk studies, SA has been demonstrated to lessen cold injury in both whole plants and just the fruits. Plant phenology may vary as a result of the exogenous administration of phytohormones like GA that may interact with other hormones. According to the findings, the exogenous administration of GA during cold stress in plants increases the countenance of the dehydration response element-binding factor (CBF1) and the GA catabolic gene *GA2ox1*. It was discovered that GA and SA working together had a synergistic impact.

4.13 CONCLUSION

Plant development and regulation in a variety of environmental situations necessitate the integration of complicated signaling pathways within the cell. *M. pipereta* is highly susceptible to environmental changes. A phytohormone is a substance that influences how plants grow and develop and aids them in overcoming numerous environmental challenges. Under stress conditions, these multifunctional chemicals drastically modify the metabolic fluctuations inside the plant cell resulting in a more resistant phenotype. Remarkably, phytohormones have been thought to be substantial stress reducers in *M. piperita* at all developmental phases, from seed germination to senescence. Abiotic stress boosts secondary metabolite concentration in *M. piperita*, which leads to a rise in medicinal properties. Under drought stress, *M. piperita* has shown a rise in the secretion of essential oil, alkaloids, and flavonoids. The possessions of salinity on *M. piperita* are highly extreme. PGRs perform important roles in various abiotic stress settings and help plants operate normally. Numerous response mechanisms incorporate signaling channels, which increase *M. piperita's* chances of survival in harsh environments. Substantial advancement has been accomplished in the investigation of the regulatory elements and hormone receptors that take part in various signaling cascades. To alleviate the detrimental effects of abiotic stress, phytohormones work in conjunction with the membrane, ROS signaling, sensors, receptors, ion channels, and MAPK signaling. Creating strains of plants that can withstand abiotic challenges might be aided by an understanding of the process of phytohormone interaction and crosstalk mediated by distinct transcriptional factors, as well as the downstream components.

KEYWORDS

- **abiotic stress**
- **peppermint**
- **phytohormones**
- **signaling mechanism**
- **crosstalk**

REFERENCES

1. Lawrence, B.M. *Mint: The Genus Mentha*. CRC Press. 2006.
2. Santoro, V.M., Nievas, F.L., Zygadlo, J.A., Giordano, W.F., Banchio, E. Effects of growth regulators on biomass and the production of secondary metabolites in peppermint (Mentha piperita) micropropagated in vitro. *American Journal of Plant Sciences*, 2013, *4*.
3. de Sousa Barros, A., de Morais, S.M., Ferreira, P.A.T., Vieira, Í.G.P., Craveiro, A.A., dos Santos Fontenelle, R.O., de Menezes, J.E.S.A., da Silva, F.W.F., de Sousa, H.A. Chemical composition and functional properties of essential oils from Mentha species. *Industrial Crops and Products*, 2015, *76*, 557–564.
4. Hasanuzzaman, M., Nahar, K., Alam, M.M., Roychowdhury, R., Fujita, M. Physiological, biochemical, and molecular mechanisms of heat stress tolerance in plants. *International Journal of Molecular Sciences*, 2013, *14*(5), 9643–9684.
5. Lata, R., Chowdhury, S., Gond, S.K., White Jr, J.F. Induction of abiotic stress tolerance in plants by endophytic microbes. *Letters in Applied Microbiology*, 2018, *66*(4), 268–276.
6. Aziz, E.E., Al-Amier, H., Craker, L.E. Influence of salt stress on growth and essential oil production in peppermint, pennyroyal, and apple mint. *Journal of Herbs, Spices & Medicinal Plants*, 2008, *14*(1–2), 77–87.
7. Khorasaninejad, S., Mousavi, A., Soltanloo, H., Hemmati, K., Khalighi, A. The effect of salinity stress on growth parameters, essential oil yield and constituent of peppermint (Mentha piperita L.). *World Applied Sciences Journal*, 2010, *11*(11), 1403–1407.
8. Baâtour, O., Tarchoune, I., Zaghdoudi, M., Chebbi, M., Ben Nasri-Ayachi, M. Essential oil and trichome density from Origanum majorana L. shoots affected by leaf age and salinity. *Bioscience Journal*, 2016, *32*(1), 238–245.
9. Ondrasek, G., Rengel, Z., Veres, S. Soil salinisation and salt stress in crop production. *Abiotic Stress in Plants—Mechanisms and Adaptations*, 2011, 171–190.
10. (10) Jayakannan, M., Bose, J., Babourina, O., Shabala, S., Massart, A., Poschenrieder, C., Rengel, Z. The NPR1-dependent salicylic acid signalling pathway is pivotal for

enhanced salt and oxidative stress tolerance in Arabidopsis. *Journal of Experimental Botany*, 2015, *66*(7), 1865–1875.

11. Azevedo, J.A., Azevedo, R.A. Heavy metals and oxidative stress: Where do we go from here. *Communications in Biometry and Crop Science*, 2006, *1*(2), 135–138.
12. Da Silva, A.J., do Nascimento, C.W.A., da Silva Gouveia-Neto, A., da Silva-Jr, E.A. LED-induced chlorophyll fluorescence spectral analysis for the early detection and monitoring of cadmium toxicity in maize plants. *Water, Air, & Soil Pollution*, 2012, *223*, 3527–3533.
13. Yang, Z., Chu, C. Towards understanding plant response to heavy metal stress. *Abiotic Stress in Plants–Mechanisms and Adaptations*, 2011, *10*, 24204.
14. Hatfield, J.L., Prueger, J.H. Temperature extremes: Effect on plant growth and development. *Weather and Climate Extremes*, 2015, *10*, 4–10.
15. Shah, N., Paulsen, G. Interaction of drought and high temperature on photosynthesis and grain-filling of wheat. *Plant and Soil*, 2003, *257*, 219–226.
16. Fahad, S., Hussain, S., Matloob, A., Khan, F.A., Khaliq, A., Saud, S., Hassan, S., Shan, D., Khan, F., Ullah, N. Phytohormones and plant responses to salinity stress: A review. *Plant Growth Regulation*, 2015, *75*, 391–404.
17. Sun, T.-P., Gubler, F. Molecular mechanism of gibberellin signaling in plants. *Annual Review of Plant Biology*, 2004, *55*, 197–223.
18. Alonso-Ramírez, A., Rodríguez, D., Reyes, D., Jiménez, J.A., Nicolás, G., López-Climent, M., Gómez-Cadenas, A., Nicolás, C. Evidence for a role of gibberellins in salicylic acid-modulated early plant responses to abiotic stress in Arabidopsis seeds. *Plant Physiology*, 2009, *150*(3), 1335–1344.
19. Gunes, A., Inal, A., Alpaslan, M., Eraslan, F., Bagci, E.G., Cicek, N. Salicylic acid induced changes on some physiological parameters symptomatic for oxidative stress and mineral nutrition in maize (Zea mays L.) grown under salinity. *Journal of Plant Physiology*, 2007, *164*(6), 728–736.
20. Horváth, E., Szalai, G., Janda, T. Induction of abiotic stress tolerance by salicylic acid signaling. *Journal of Plant Growth Regulation*, 2007, *26*, 290–300.
21. Nazar, R., Iqbal, N., Syeed, S., Khan, N.A. Salicylic acid alleviates decreases in photosynthesis under salt stress by enhancing nitrogen and sulfur assimilation and antioxidant metabolism differentially in two mungbean cultivars. *Journal of Plant Physiology*, 2011, *168*(8), 807–815.
22. Palma, F., López-Gómez, M., Tejera, N.A., Lluch, C. Salicylic acid improves the salinity tolerance of Medicago sativa in symbiosis with Sinorhizobium meliloti by preventing nitrogen fixation inhibition. *Plant Science*, 2013, *208*, 75–82.
23. Raskin, I. Role of salicylic acid in plants. *Annual Review of Plant Biology*, 1992, *43*(1), 439–463.
24. Aldesuquy, H., Mankarios, A., Awad, H. Effect of some antitranspirants on growth, metabolism and productivity of saline-treated wheat plants. Induction of stomatal closure, inhibition of transpiration and improvement of leaf turgidity. *Acta Botanica Hungarica*, 1998, *41*, 1–10.
25. Rajasekaran, L., Blake, T. New plant growth regulators protect photosynthesis and enhance growth under drought of jack pine seedlings. *Journal of Plant Growth Regulation*, 1999, *18*, 175–181.

26. Srivastava, M.K., Dwivedi, U.N. Delayed ripening of banana fruit by salicylic acid. *Plant Science*, 2000, *158*(1–2), 87–96.
27. Upreti, K., Sharma, M. Role of plant growth regulators in abiotic stress tolerance. *Abiotic Stress Physiology of Horticultural Crops*, 2016, 19–46.
28. Vishwakarma, K., Upadhyay, N., Kumar, N., Yadav, G., Singh, J., Mishra, R.K., Kumar, V., Verma, R., Upadhyay, R., Pandey, M. Abscisic acid signaling and abiotic stress tolerance in plants: A review on current knowledge and future prospects. *Frontiers in Plant Science*, 2017, *8*, 161.
29. Salvi, P., Manna, M., Kaur, H., Thakur, T., Gandass, N., Bhatt, D., Muthamilarasan, M. Phytohormone signaling and crosstalk in regulating drought stress response in plants. *Plant Cell Reports*, 2021, *40*, 1305–1329.
30. O'Brien, J.A., Benková, E. Cytokinin cross-talking during biotic and abiotic stress responses. *Frontiers in Plant Science*, 2013, *4*, 451.
31. Sah, S.K., Reddy, K.R., Li, J. Abscisic acid and abiotic stress tolerance in crop plants. *Frontiers in Plant Science*, 2016, *7*, 571.
32. Puértolas, J., Conesa, M.R., Ballester, C., Dodd, I.C. Local root abscisic acid (ABA) accumulation depends on the spatial distribution of soil moisture in potato: Implications for ABA signalling under heterogeneous soil drying. *Journal of Experimental Botany*, 2015, *66*(8), 2325–2334.
33. Mehrotra, R., Jain, V., Shekhar, C., Mehrotra, S. Genome wide analysis of Arabidopsis thaliana reveals high frequency of AAAGN7CTTT motif. *Meta Gene*, 2014, *2*, 606–615.
34. Xiong, L., Schumaker, K.S., Zhu, J.-K. Cell signaling during cold, drought, and salt stress. *The Plant Cell*, 2002, *14*(suppl_1), S165–S183.
35. Xiong, L., Ishitani, M., Lee, H., Zhu, J.-K. The Arabidopsis LOS5/ABA3 locus encodes a molybdenum cofactor sulfurase and modulates cold stress–and osmotic stress–responsive gene expression. *The Plant Cell*, 2001, *13*(9), 2063–2083.
36. Thomashow, M.F. Plant cold acclimation: Freezing tolerance genes and regulatory mechanisms. *Annual Review of Plant Biology*, 1999, *50*(1), 571–599.
37. Shinozaki, K., Yamaguchi-Shinozaki, K. Molecular responses to dehydration and low temperature: Differences and cross-talk between two stress signaling pathways. *Current Opinion in Plant Biology*, 2000, *3*(3), 217–223.
38. Uno, Y., Furihata, T., Abe, H., Yoshida, R., Shinozaki, K., Yamaguchi-Shinozaki, K. Arabidopsis basic leucine zipper transcription factors involved in an abscisic acid-dependent signal transduction pathway under drought and high-salinity conditions. *Proceedings of the National Academy of Sciences*, 2000, *97*(21), 11632–11637.
39. Knight, H., Trewavas, A.J., Knight, M.R. Calcium signalling in Arabidopsis thaliana responding to drought and salinity. *The Plant Journal*, 1997, *12*(5), 1067–1078.
40. Yamaguchi-Shinozaki, K., Shinozaki, K. Characterization of the expression of a desiccation-responsive rd29 gene of Arabidopsis thaliana and analysis of its promoter in transgenic plants. *Molecular and General Genetics MGG*, 1993, *236*, 331–340.
41. Mahajan, S., Sopory, S.K., Tuteja, N. CBL-CIPK paradigm: Role in calcium and stress signaling in plants. *Proceedings of the Indian National Science Academy*, 2006, *72*(2), 63.

42. Sanan-Mishra, N., Pham, X.H., Sopory, S.K., Tuteja, N. Pea DNA helicase 45 overexpression in tobacco confers high salinity tolerance without affecting yield. *Proceedings of the National Academy of Sciences*, 2005, *102*(2), 509–514.
43. Lee, M., Jung, J.-H., Han, D.-Y., Seo, P.J., Park, W.J., Park, C.-M. Activation of a flavin monooxygenase gene YUCCA7 enhances drought resistance in Arabidopsis. *Planta*, 2012, *235*, 923–938.
44. Saini, S., Kaur, N., Pati, P.K. Phytohormones: Key players in the modulation of heavy metal stress tolerance in plants. *Ecotoxicology and Environmental Safety*, 2021, *223*, 112578.
45. Lytvyn, D.I., Raynaud, C., Yemets, A.I., Bergounioux, C., Blume, Y.B. Involvement of inositol biosynthesis and nitric oxide in the mediation of UV-B induced oxidative stress. *Frontiers in Plant Science*, 2016, *7*, 430.
46. Lavy, M., Estelle, M. Mechanisms of auxin signaling. *Development*, 2016, *143*(18), 3226–3229.
47. Liu, J., Moore, S., Chen, C., Lindsey, K. Crosstalk complexities between auxin, cytokinin, and ethylene in Arabidopsis root development: From experiments to systems modeling, and back again. *Molecular Plant*, 2017, *10*(12), 1480–1496.
48. Sharma, E., Sharma, R., Borah, P., Jain, M., Khurana, J.P. Emerging roles of auxin in abiotic stress responses. *Elucidation of Abiotic Stress Signaling in Plants: Functional Genomics Perspectives*, 2015, *1*, 299–328.
49. Morkunas, I., Mai, V.C., Waśkiewicz, A., Formela, M., Goliński, P. Major phytohormones under abiotic stress. *Physiological Mechanisms and Adaptation Strategies in Plants Under Changing Environment*, 2014, *2*, 87–135.
50. Seidel, C., Walz, A., Park, S., Cohen, J., Ludwig-Müller, J. Indole-3-acetic acid protein conjugates: Novel players in auxin homeostasis. *Plant Biology*, 2006, *8*(3), 340–345.
51. Normanly, J. Approaching cellular and molecular resolution of auxin biosynthesis and metabolism. *Cold Spring Harbor Perspectives in Biology*, 2010, *2*(1), a001594.
52. Hagen, G., Guilfoyle, T. Auxin-responsive gene expression: Genes, promoters and regulatory factors. *Plant Molecular Biology*, 2002, *49*, 373–385.
53. Woodward, A.W., Bartel, B. Auxin: Regulation, action, and interaction. *Annals of Botany*, 2005, *95*(5), 707–735.
54. Shibasaki, K., Uemura, M., Tsurumi, S., Rahman, A. Auxin response in Arabidopsis under cold stress: Underlying molecular mechanisms. *The Plant Cell*, 2009, *21*(12), 3823–3838.
55. Du, H., Wu, N., Fu, J., Wang, S., Li, X., Xiao, J., Xiong, L. A GH3 family member, OsGH3-2, modulates auxin and abscisic acid levels and differentially affects drought and cold tolerance in rice. *Journal of Experimental Botany*, 2012, *63*(18), 6467–6480.
56. Du, H., Wu, N., Chang, Y., Li, X., Xiao, J., Xiong, L. Carotenoid deficiency impairs ABA and IAA biosynthesis and differentially affects drought and cold tolerance in rice. *Plant Molecular Biology*, 2013, *83*, 475–488.
57. Im Kim, J., Baek, D., Park, H.C., Chun, H.J., Oh, D.-H., Lee, M.K., Cha, J.-Y., Kim, W.-Y., Kim, M.C., Chung, W.S. Overexpression of Arabidopsis YUCCA6 in potato results in high-auxin developmental phenotypes and enhanced resistance to water deficit. *Molecular Plant*, 2013, *6*(2), 337–349.

58. Bhandari, S., Nailwal, T.K. Role of brassinosteroids in mitigating abiotic stresses in plants. *Biologia*, 2020, *75*(12), 2203–2230.
59. Alhaithloul, H.A.S., Soliman, M.H. Methyl jasmonate and brassinosteroids: Emerging plant growth regulators in plant abiotic stress tolerance and environmental changes. In: *Plant Growth Regulators: Signalling Under Stress Conditions*. 2021; pp. 173–195.
60. Karlidag, H., Yildirim, E., Turan, M. Role of 24-epibrassinolide in mitigating the adverse effects of salt stress on stomatal conductance, membrane permeability, and leaf water content, ionic composition in salt stressed strawberry (Fragaria× ananassa). *Scientia Horticulturae*, 2011, *130*(1), 133–140.
61. Friedrichsen, D.M., Joazeiro, C.A., Li, J., Hunter, T., Chory, J. Brassinosteroid-insensitive-1 is a ubiquitously expressed leucine-rich repeat receptor serine/threonine kinase. *Plant Physiology*, 2000, *123*(4), 1247–1256.
62. Kinoshita, T., Caño-Delgado, A., Seto, H., Hiranuma, S., Fujioka, S., Yoshida, S., Chory, J. Binding of brassinosteroids to the extracellular domain of plant receptor kinase BRI1. *Nature*, 2005, *433*(7022), 167–171.
63. Choudhary, S.P., Yu, J.-Q., Yamaguchi-Shinozaki, K., Shinozaki, K., Tran, L.-S.P. Benefits of brassinosteroid crosstalk. *Trends in Plant Science*, 2012, *17*(10), 594–605.
64. Wang, Z.-Y., Wang, Q., Chong, K., Wang, F., Wang, L., Bai, M., Jia, C. The brassinosteroid signal transduction pathway. *Cell Research*, 2006, *16*(5), 427–434.
65. Kim, T.-W., Wang, Z.-Y. Brassinosteroid signal transduction from receptor kinases to transcription factors. *Annual Review of Plant Biology*, 2010, *61*, 681–704.
66. Fahad, S., Nie, L., Chen, Y., Wu, C., Xiong, D., Saud, S., Hongyan, L., Cui, K., Huang, J. Crop plant hormones and environmental stress. *Sustainable Agriculture Reviews*, 2015, *15*, 371–400.
67. Cortleven, A., Ehret, S., Schmülling, T., Johansson, H. Ethylene-independent promotion of photomorphogenesis in the dark by cytokinin requires COP1 and the CDD complex. *Journal of Experimental Botany*, 2019, *70*(1), 165–178.
68. Zwack, P.J., Rashotte, A.M. Interactions between cytokinin signalling and abiotic stress responses. *Journal of Experimental Botany*, 2015, *66*(16), 4863–4871.
69. Ha, S., Vankova, R., Yamaguchi-Shinozaki, K., Shinozaki, K., Tran, L.-S.P. Cytokinins: Metabolism and function in plant adaptation to environmental stresses. *Trends in Plant Science*, 2012, *17*(3), 172–179.
70. Rasool, S., Urwat, U., Nazir, M., Zargar, S.M., Zargar, M. Cross talk between phytohormone signaling pathways under abiotic stress conditions and their metabolic engineering for conferring abiotic stress tolerance. In: *Abiotic stress-Mediated Sensing and Signaling in Plants: An Omics Perspective*. 2018; pp. 329–350.
71. Alhaithloul, H.A.S., Abu-Elsaoud, A.M., Soliman, M.H. Abiotic stress tolerance in crop plants: Role of phytohormones. In: *Abiotic Stress in Plants*. 2020; p. 233.
72. Schaller, G.E. Ethylene and the regulation of plant development. *BMC Biology*, 2012, *10*(1), 1–3.
73. Gamalero, E., Glick, B.R. Ethylene and abiotic stress tolerance in plants. In: *Environmental Adaptations and Stress Tolerance of Plants in the Era of Climate Change*. 2012; pp. 395–412.
74. Klay, I., Pirrello, J., Riahi, L., Bernadac, A., Cherif, A., Bouzayen, M., Bouzid, S. Ethylene response factor Sl-ERF. B. 3 is responsive to abiotic stresses and mediates

salt and cold stress response regulation in tomato. *The Scientific World Journal*, 2014, *2014*, 167681.

75. Tsuchisaka, A., Theologis, A. Heterodimeric interactions among the 1-amino-cyclopropane-1-carboxylate synthase polypeptides encoded by the Arabidopsis gene family. *Proceedings of the National Academy of Sciences*, 2004, *101*(8), 2275–2280.
76. Penninckx, I.A., Thomma, B.P., Buchala, A., Métraux, J.-P., Broekaert, W.F. Concomitant activation of jasmonate and ethylene response pathways is required for induction of a plant defensin gene in Arabidopsis. *The Plant Cell*, 1998, *10*(12), 2103–2113.
77. Apelbaum, A., Burgoon, A.C., Anderson, J.D., Solomos, T., Lieberman, M. Some characteristics of the system converting 1-aminocyclopropane-1-carboxylic acid to ethylene. *Plant Physiology*, 1981, *67*(1), 80–84.
78. McKeon, T.A., Hoffman, N.E., Yang, S.F. The effect of plant-hormone pretreatments on ethylene production and synthesis of 1-aminocyclopropane-1-carboxylic acid in water-stressed wheat leaves. *Planta*, 1982, *155*, 437–443.
79. Tanaka, Y., Sano, T., Tamaoki, M., Nakajima, N., Kondo, N., Hasezawa, S. Cytokinin and auxin inhibit abscisic acid-induced stomatal closure by enhancing ethylene production in Arabidopsis. *Journal of Experimental Botany*, 2006, *57*(10), 2259–2266.
80. Zhao, X.-C., Schaller, G.E. Effect of salt and osmotic stress upon expression of the ethylene receptor ETR1 in Arabidopsis thaliana. *Febs Letters*, 2004, *562*(1–3), 189–192.
81. Peng, H.-P., Lin, T.-Y., Wang, N.-N., Shih, M.-C. Differential expression of genes encoding 1-aminocyclopropane-1-carboxylate synthase in Arabidopsis during hypoxia. *Plant Molecular Biology*, 2005, *58*, 15–25.
82. Mehlhorn, H., Wellburn, A.R. Stress ethylene formation determines plant sensitivity to ozone. *Nature*, 1987, *327*(6121), 417–418.
83. Overmyer, K., Brosché, M., Kangasjärvi, J. Reactive oxygen species and hormonal control of cell death. *Trends in Plant Science*, 2003, *8*(7), 335–342.
84. Ogawa, D., Nakajima, N., Sano, T., Tamaoki, M., Aono, M., Kubo, A., Kanna, M., Ioki, M., Kamada, H., Saji, H. Salicylic acid accumulation under O3 exposure is regulated by ethylene in tobacco plants. *Plant and Cell Physiology*, 2005, *46*(7), 1062–1072.
85. Tuominen, H., Overmyer, K., KeinaÈnen, M., Kollist, H., Kangasjärvi, J. Mutual antagonism of ethylene and jasmonic acid regulates ozone-induced spreading cell death in Arabidopsis. *The Plant Journal*, 2004, *39*(1), 59–69.
86. Nakajima, N., Itoh, T., Takikawa, S., Asai, N., Tamaoki, M., Aono, M., Kubo, A., Azumi, Y., Kamada, H., Saji, H. Improvement in ozone tolerance of tobacco plants with an antisense DNA for 1-aminocyclopropane-1-carboxylate synthase. *Plant, Cell & Environment*, 2002, *25*(6), 727–735.
87. Sinn, J.P., Schlagnhaufer, C.D., Arteca, R.N., Pell, E.J. Ozone-induced ethylene and foliar injury responses are altered in 1-aminocyclopropane-1-carboxylate synthase antisense potato plants. *New Phytologist*, 2004, *164*(2), 267–277.
88. Iqbal, N., Nazar, R., Khan, M.I.R., Masood, A., Khan, N.A. Role of gibberellins in regulation of source-sink relations under optimal and limiting environmental conditions. *Current Science (Bangalore)*, 2011, *100*(7), 998–1007.

89. Claeys, H., Skirycz, A., Maleux, K., Inzé, D. DELLA signaling mediates stress-induced cell differentiation in Arabidopsis leaves through modulation of anaphase-promoting complex/cyclosome activity. *Plant Physiology*, 2012, *159*(2), 739–747.
90. Gómez-Cadenas, A., de Ollas, C., Manzi, M., Arbona, V. Phytohormonal crosstalk under abiotic stress. In: *Phytohormones: A Window to Metabolism, Signaling and Biotechnological Applications*. 2014; pp. 289–321.
91. Achard, P., Cheng, H., De Grauwe, L., Decat, J., Schoutteten, H., Moritz, T., Van Der Straeten, D., Peng, J., Harberd, N.P. Integration of plant responses to environmentally activated phytohormonal signals. *Science*, 2006, *311*(5757), 91–94.
92. Achard, P., Gong, F., Cheminant, S., Alioua, M., Hedden, P., Genschik, P. The cold-inducible CBF1 factor-dependent signaling pathway modulates the accumulation of the growth-repressing DELLA proteins via its effect on gibberellin metabolism. *The Plant Cell*, 2008, *20*(8), 2117–2129.
93. Magome, H., Yamaguchi, S., Hanada, A., Kamiya, Y., Oda, K. The DDF1 transcriptional activator upregulates expression of a gibberellin-deactivating gene, GA2ox7, under high-salinity stress in Arabidopsis. *The Plant Journal*, 2008, *56*(4), 613–626.
94. Magome, H., Yamaguchi, S., Hanada, A., Kamiya, Y., Oda, K. dwarf and delayed-flowering 1, a novel Arabidopsis mutant deficient in gibberellin biosynthesis because of overexpression of a putative AP2 transcription factor. *The Plant Journal*, 2004, *37*(5), 720–729.
95. Achard, P., Gusti, A., Cheminant, S., Alioua, M., Dhondt, S., Coppens, F., Beemster, G.T., Genschik, P. Gibberellin signaling controls cell proliferation rate in Arabidopsis. *Current Biology*, 2009, *19*(14), 1188–1193.
96. Fukao, T., Yeung, E., Bailey-Serres, J. The submergence tolerance regulator SUB1A mediates crosstalk between submergence and drought tolerance in rice. *The Plant Cell*, 2011, *23*(1), 412–427.
97. Fukao, T., Bailey-Serres, J. Submergence tolerance conferred by Sub1A is mediated by SLR1 and SLRL1 restriction of gibberellin responses in rice. *Proceedings of the National Academy of Sciences*, 2008, *105*(43), 16814–16819.
98. Ubeda-Tomás, S., Federici, F., Casimiro, I., Beemster, G.T., Bhalerao, R., Swarup, R., Doerner, P., Haseloff, J., Bennett, M.J. Gibberellin signaling in the endodermis controls Arabidopsis root meristem size. *Current Biology*, 2009, *19*(14), 1194–1199.
99. Viswanath, K.K., Varakumar, P., Pamuru, R.R., Basha, S.J., Mehta, S., Rao, A.D. Plant lipoxygenases and their role in plant physiology. *Journal of Plant Biology*, 2020, *63*, 83–95.
100. Dar, T.A., Uddin, M., Khan, M.M.A., Hakeem, K., Jaleel, H. Jasmonates counter plant stress: A review. *Environmental and Experimental Botany*, 2015, *115*, 49–57.
101. Abouelsaad, I., Renault, S. Enhanced oxidative stress in the jasmonic acid-deficient tomato mutant def-1 exposed to NaCl stress. *Journal of Plant Physiology*, 2018, *226*, 136–144.
102. Alisofi, S., Einali, A., Sangtarash, M.H. Jasmonic acid-induced metabolic responses in bitter melon (Momordica charantia) seedlings under salt stress. *The Journal of Horticultural Science and Biotechnology*, 2020, *95*(2), 247–259.
103. Bali, S., Kaur, P., Kohli, S.K., Ohri, P., Thukral, A.K., Bhardwaj, R., Wijaya, L., Alyemeni, M.N., Ahmad, P. Jasmonic acid induced changes in physio-biochemical

attributes and ascorbate-glutathione pathway in Lycopersicon esculentum under lead stress at different growth stages. *Science of the Total Environment*, 2018, *645*, 1344–1360.
104. Zhao, Y., Song, C., Brummell, D.A., Qi, S., Lin, Q., Bi, J., Duan, Y. Salicylic acid treatment mitigates chilling injury in peach fruit by regulation of sucrose metabolism and soluble sugar content. *Food Chemistry*, 2021, *358*, 129867.
105. Zhao, S., Ma, Q., Xu, X., Li, G., Hao, L. Tomato jasmonic acid-deficient mutant spr2 seedling response to cadmium stress. *Journal of Plant Growth Regulation*, 2016, *35*, 603–610.
106. Meng, X., Han, J., Wang, Q., Tian, S. Changes in physiology and quality of peach fruits treated by methyl jasmonate under low temperature stress. *Food Chemistry*, 2009, *114*(3), 1028–1035.
107. Krasuska, U., Ciacka, K., Gniazdowska, A. Nitric oxide-polyamines cross-talk during dormancy release and germination of apple embryos. *Nitric Oxide*, 2017, *68*, 38–50.
108. Desikan, R., A.-H.-Mackerness, S., Hancock, J.T., Neill, S.J. Regulation of the Arabidopsis transcriptome by oxidative stress. *Plant Physiology*, 2001, *127*(1), 159–172.
109. Garcıa-Mata, C., Lamattina, L. Nitric oxide and abscisic acid cross talk in guard cells. *Plant Physiology*, 2002, *128*(3), 790–792.
110. Tossi, V., Lamattina, L., Jenkins, G., Cassia, R. UV-B-induced stomatal closure in Arabidopsis is regulated by the UVR8 photoreceptor in an NO-dependent mechanism. *Plant Physiology*, 2014, *164*(4), 2220–2230.
111. Wang, Y., Loake, G.J., Chu, C. Cross-talk of nitric oxide and reactive oxygen species in plant programed cell death. *Frontiers in Plant Science*, 2013, *4*, 314.
112. Asgher, M., Per, T.S., Masood, A., Fatma, M., Freschi, L., Corpas, F.J., Khan, N.A. Nitric oxide signaling and its crosstalk with other plant growth regulators in plant responses to abiotic stress. *Environmental Science and Pollution Research*, 2017, *24*, 2273–2285.
113. Khan, M.I.R., Nazir, F., Asgher, M., Per, T.S., Khan, N.A. Selenium and sulfur influence ethylene formation and alleviate cadmium-induced oxidative stress by improving proline and glutathione production in wheat. *Journal of Plant Physiology*, 2015, *173*, 9–18.
114. Fancy, N.N., Bahlmann, A.K., Loake, G.J. Nitric oxide function in plant abiotic stress. *Plant, Cell & Environment*, 2017, *40*(4), 462–472.
115. Li, Z., Luo, L., Sun, Y. Signal crosstalk between nitric oxide and hydrogen sulfide may be involved in hydrogen peroxide-induced thermotolerance in maize seedlings. *Russian Journal of Plant Physiology*, 2015, *62*, 507–514.
116. Shan, C., Zhou, Y., Liu, M. Nitric oxide participates in the regulation of the ascorbate-glutathione cycle by exogenous jasmonic acid in the leaves of wheat seedlings under drought stress. *Protoplasma*, 2015, *252*, 1397–1405.
117. Santos, M., Zandonadi, D., de Sá, A. Abscisic acidnitric oxide and auxin interaction modulates salt stress response in tomato roots. *Theoretical and Experimental Plant Physiology*, 2020, *32*, 301–313.
118. Taylor, J.E., McAinsh, M.R. Signalling crosstalk in plants: Emerging issues. *Journal of Experimental Botany*, 2004, *55*(395), 147–149.

119. Peck, S., Mittler, R. Plant signaling in biotic and abiotic stress. *Journal of Experimental Botany*, 2020, *71*, 1649–1651.
120. Falak, N., Imran, Q.M., Hussain, A., Yun, B.-W. Transcription factors as the "Blitzkrieg" of plant defense: A pragmatic view of nitric oxide's role in gene regulation. *International Journal of Molecular Sciences*, 2021, *22*(2), 522.
121. Sánchez-Vicente, I., Fernández-Espinosa, M.G., Lorenzo, O. Nitric oxide molecular targets: Reprogramming plant development upon stress. *Journal of Experimental Botany*, 2019, *70*(17), 4441–4460.
122. Sami, F., Faizan, M., Faraz, A., Siddiqui, H., Yusuf, M., Hayat, S. Nitric oxide-mediated integrative alterations in plant metabolism to confer abiotic stress tolerance, NO crosstalk with phytohormones and NO-mediated post translational modifications in modulating diverse plant stress. *Nitric Oxide*, 2018, *73*, 22–38.
123. Ma, Z., Marsolais, F., Bykova, N.V., Igamberdiev, A.U. Nitric oxide and reactive oxygen species mediate metabolic changes in barley seed embryo during germination. *Frontiers in Plant Science*, 2016, *7*, 138.
124. Gaupels, F., Furch, A.C., Zimmermann, M.R., Chen, F., Kaever, V., Buhtz, A., Kehr, J., Sarioglu, H., Kogel, K.-H., Durner, J. Systemic induction of NO-, redox-, and cGMP signaling in the pumpkin extrafascicular phloem upon local leaf wounding. *Frontiers in Plant Science*, 2016, *7*, 154.
125. Singhal, R.K., Jatav, H.S., Aftab, T., Pandey, S., Mishra, U.N., Chauhan, J., Chand, S., Indu, Saha, D., Dadarwal, B.K. Roles of nitric oxide in conferring multiple abiotic stress tolerance in plants and crosstalk with other plant growth regulators. *Journal of Plant Growth Regulation*, 2021, *40*, 2303–2328.
126. Sivakumaran, A., Akinyemi, A., Mandon, J., Cristescu, S.M., Hall, M.A., Harren, F.J., Mur, L.A. ABA suppresses Botrytis cinerea elicited NO production in tomato to influence H2O2 generation and increase host susceptibility. *Frontiers in Plant Science*, 2016, *7*, 709.
127. Gupta, K.J., Igamberdiev, A.U., Manjunatha, G., Segu, S., Moran, J.F., Neelawarne, B., Bauwe, H., Kaiser, W.M. The emerging roles of nitric oxide (NO) in plant mitochondria. *Plant Science*, 2011, *181*(5), 520–526.
128. Begara-Morales, J.C., Sánchez-Calvo, B., Chaki, M., Valderrama, R., Mata-Pérez, C., Padilla, M.N., Corpas, F.J., Barroso, J.B. Antioxidant systems are regulated by nitric oxide-mediated post-translational modifications (NO-PTMs). *Frontiers in Plant Science*, 2016, *7*, 152.
129. Romero-Puertas, M.C., Sandalio, L.M. Nitric oxide level is self-regulating and also regulates its ROS partners. *Frontiers Media SA*, 2016, *7*, 316.
130. Foyer, C.H., Noctor, G. Defining robust redox signalling within the context of the plant cell. *Wiley Online Library*, 2015, *38*, 239–239.
131. Yang, J., Deng, X., Wang, X., Wang, J., Du, S., Li, Y. The calcium sensor OsCBL1 modulates nitrate signaling to regulate seedling growth in rice. *PLoS One*, 2019, *14*(11), e0224962.
132. Gross, I., Durner, J. In search of enzymes with a role in 3′, 5′-cyclic guanosine monophosphate metabolism in plants. *Frontiers in Plant Science*, 2016, *7*, 576.
133. Simon, R., Dresselhaus, T. Peptides take centre stage in plant signalling. *Journal of Experimental Botany*, 2015, *66*(17), 5135–5138.

134. Fatma, M., Masood, A., Per, T.S., Khan, N.A. Nitric oxide alleviates salt stress inhibited photosynthetic performance by interacting with sulfur assimilation in mustard. *Frontiers in Plant Science*, 2016, *7*, 521.
135. Wu, H., Zheng, Y., Liu, J., Zhang, H., Chen, H. Heme oxygenase-1 delays gibberellin-induced programmed cell death of rice aleurone layers subjected to drought stress by interacting with nitric oxide. *Frontiers in Plant Science*, 2016, *6*, 1267.
136. Groppa, M., Benavides, M. Polyamines and abiotic stress: Recent advances. *Amino Acids*, 2008, *34*, 35–45.
137. Chen, D., Shao, Q., Yin, L., Younis, A., Zheng, B. Polyamine function in plants: Metabolism, regulation on development, and roles in abiotic stress responses. *Frontiers in Plant Science*, 2019, *9*, 1945.
138. Hussain, S.S., Ali, M., Ahmad, M., Siddique, K.H. Polyamines: Natural and engineered abiotic and biotic stress tolerance in plants. *Biotechnology Advances*, 2011, *29*(3), 300–311.
139. Choudhary, S.P., Kanwar, M., Bhardwaj, R., Yu, J.-Q., Tran, L.-S.P. Chromium stress mitigation by polyamine-brassinosteroid application involves phytohormonal and physiological strategies in Raphanus sativus L. *PLoS One*, 2012, *7*(3), e33210.
140. Hassan, F., Ali, E., Alamer, K. Exogenous application of polyamines alleviates water stress-induced oxidative stress of Rosa damascena Miller var. trigintipetala Dieck. *South African Journal of Botany*, 2018, *116*, 96–102.
141. Kapulnik, Y., Koltai, H. Strigolactone involvement in root development, response to abiotic stress, and interactions with the biotic soil environment. *Plant Physiology*, 2014, *166*(2), 560–569.
142. van Zeijl, A., Liu, W., Xiao, T.T., Kohlen, W., Yang, W.-C., Bisseling, T., Geurts, R. The strigolactone biosynthesis gene DWARF27 is co-opted in rhizobium symbiosis. *BMC Plant Biology*, 2015, *15*(1), 1–15.
143. Pandey, A., Sharma, M., Pandey, G.K. Emerging roles of strigolactones in plant responses to stress and development. *Frontiers in Plant Science*, 2016, *7*, 434.
144. Min, Z., Li, R., Chen, L., Zhang, Y., Li, Z., Liu, M., Ju, Y., Fang, Y. Alleviation of drought stress in grapevine by foliar-applied strigolactones. *Plant Physiology and Biochemistry*, 2019, *135*, 99–110.
145. Santoro, V., Schiavon, M., Gresta, F., Ertani, A., Cardinale, F., Sturrock, C.J., Celi, L., Schubert, A. Strigolactones control root system architecture and tip anatomy in Solanum lycopersicum L. plants under P starvation. *Plants*, 2020, *9*(5), 612.
146. Visentin, I., Vitali, M., Ferrero, M., Zhang, Y., Ruyter-Spira, C., Novák, O., Strnad, M., Lovisolo, C., Schubert, A., Cardinale, F. Low levels of strigolactones in roots as a component of the systemic signal of drought stress in tomato. *New Phytologist*, 2016, *212*(4), 954–963.
147. Bhoi, A., Yadu, B., Chandra, J., Keshavkant, S. Contribution of strigolactone in plant physiology, hormonal interaction and abiotic stresses. *Planta*, 2021, *254*, 1–21.
148. Khan, N., Bano, A., Ali, S., Babar, M.A. Crosstalk amongst phytohormones from planta and PGPR under biotic and abiotic stresses. *Plant Growth Regulation*, 2020, *90*, 189–203.
149. Jogawat, A., Yadav, B., Chhaya, Lakra, N., Singh, A.K., Narayan, O.P. Crosstalk between phytohormones and secondary metabolites in the drought stress tolerance of crop plants: A review. *Physiologia Plantarum*, 2021, *172*(2), 1106–1132.

150. Wani, S.H., Kumar, V., Shriram, V., Sah, S.K. Phytohormones and their metabolic engineering for abiotic stress tolerance in crop plants. *The Crop Journal*, 2016, *4*(3), 162–176.
151. Niharika, Singh, N.B., Singh, A., Khare, S., Yadav, V., Bano, C., Yadav, R.K. Mitigating strategies of gibberellins in various environmental cues and their crosstalk with other hormonal pathways in plants: A review. *Plant Molecular Biology Reporter*, 2021, *39*, 34–49.
152. Xing, Q., Zhang, X., Li, Y., Shao, Q., Cao, S., Wang, F., Qi, H. The lipoxygenase CmLOX13 from oriental melon enhanced severe drought tolerance via regulating ABA accumulation and stomatal closure in Arabidopsis. *Environmental and Experimental Botany*, 2019, *167*, 103815.
153. Kang, C., He, S., Zhai, H., Li, R., Zhao, N., Liu, Q. A sweetpotato auxin response factor gene (IbARF5) is involved in carotenoid biosynthesis and salt and drought tolerance in transgenic Arabidopsis. *Frontiers in Plant Science*, 2018, *9*, 1307.
154. Zhang, X.-W., Liu, F.-J., Zhai, J., Li, F.-D., Bi, H.-G., Ai, X.-Z. Auxin acts as a downstream signaling molecule involved in hydrogen sulfide-induced chilling tolerance in cucumber. *Planta*, 2020, *251*, 1–19.
155. Liu, Q., Gao, T., Liu, W., Liu, Y., Zhao, Y., Liu, Y., Li, W., Ding, K., Ma, F., Li, C. Functions of dopamine in plants: A review. *Plant Signaling & Behavior*, 2020, *15*(12), 1827782.
156. Kulma, A., Szopa, J. Catecholamines are active compounds in plants. *Plant Science*, 2007, *172*(3), 433–440.
157. Müller, M. Foes or friends: ABA and ethylene interaction under abiotic stress. *Plants*, 2021, *10*(3), 448.
158. He, R., Zhuang, Y., Cai, Y., Agüero, C.B., Liu, S., Wu, J., Deng, S., Walker, M.A., Lu, J., Zhang, Y. Overexpression of 9-cis-epoxycarotenoid dioxygenase cisgene in grapevine increases drought tolerance and results in pleiotropic effects. *Frontiers in Plant Science*, 2018, *9*, 970.
159. Altaf, M.A., Shahid, R., Kumar, R., Altaf, M.M., Kumar, A., Khan, L.U., Saqib, M., Nawaz, M.A., Saddiq, B., Bahadur, S. Phytohormones mediated modulation of abiotic stress tolerance and potential crosstalk in horticultural crops. *Journal of Plant Growth Regulation*, 2023, *42*(8), 4724–4750.
160. Dayan, J., Schwarzkopf, M., Avni, A., Aloni, R. Enhancing plant growth and fiber production by silencing GA 2-oxidase. *Plant Biotechnology Journal*, 2010, *8*(4), 425–435.
161. Shi, J., Wang, J., Wang, N., Zhou, H., Xu, Q., Yan, G. Overexpression of StGA2ox1 gene increases the tolerance to abiotic stress in transgenic potato (Solanum tuberosum L.) plants. *Applied Biochemistry and Biotechnology*, 2019, *187*(4), 1204–1219.
162. Du, P., Yin, B., Cao, Y., Han, R., Ji, J., He, X., Liang, B., Xu, J. Beneficial effects of exogenous melatonin and dopamine on low nitrate stress in Malus hupehensis. *Frontiers in Plant Science*, 2022, *12*, 807472.
163. Per, T.S., Khan, M.I.R., Anjum, N.A., Masood, A., Hussain, S.J., Khan, N.A. Jasmonates in plants under abiotic stresses: Crosstalk with other phytohormones matters. *Environmental and Experimental Botany*, 2018, *145*, 104–120.
164. Jyoti, S.D., Azim, J.B., Robin, A.H.K. Genome-wide characterization and expression profiling of EIN3/EIL family genes in Zea mays. *Plant Gene*, 2021, *25*, 100270.

165. Liu, C., Li, J., Zhu, P., Yu, J., Hou, J., Wang, C., Long, D., Yu, M., Zhao, A. Mulberry EIL3 confers salt and drought tolerances and modulates ethylene biosynthetic gene expression. *PeerJ*, 2019, *7*, e6391.
166. Protacio, C.M., Dai, Y.-R., Lewis, E.F., Flores, H.E. Growth stimulation by catecholamines in plant tissue/organ cultures. *Plant Physiology*, 1992, *98*(1), 89–96.
167. Liang, B., Ma, C., Zhang, Z., Wei, Z., Gao, T., Zhao, Q., Ma, F., Li, C. Long-term exogenous application of melatonin improves nutrient uptake fluxes in apple plants under moderate drought stress. *Environmental and Experimental Botany*, 2018, *155*, 650–661.
168. Oh, E., Zhu, J.-Y., Bai, M.-Y., Arenhart, R.A., Sun, Y., Wang, Z.-Y. Cell elongation is regulated through a central circuit of interacting transcription factors in the Arabidopsis hypocotyl. *Elife*, 2014, *3*, e03031.
169. Shoresh, M., Spivak, M., Bernstein, N. Involvement of calcium-mediated effects on ROS metabolism in the regulation of growth improvement under salinity. *Free Radical Biology and Medicine*, 2011, *51*(6), 1221–1234.
170. Bitrián, M., Zarza, X., Altabella, T., Tiburcio, A.F., Alcázar, R. Polyamines under abiotic stress: Metabolic crossroads and hormonal crosstalks in plants. *Metabolites*, 2012, *2*(3), 516–528.
171. Anwar, R., Mattoo, A.K., Handa, A.K. Polyamine interactions with plant hormones: Crosstalk at several levels. In: *Polyamines: A Universal Molecular Nexus for Growth, Survival, and Specialized Metabolism*. 2015; pp. 267–302.
172. Mukherjee, S. Novel perspectives on the molecular crosstalk mechanisms of serotonin and melatonin in plants. *Plant Physiology and Biochemistry*, 2018, *132*, 33–45.
173. Caarls, L., Pieterse, C.M., Van Wees, S.C. How salicylic acid takes transcriptional control over jasmonic acid signaling. *Frontiers in Plant Science*, 2015, *6*, 170.
174. Navarro, L., Bari, R., Achard, P., Lisón, P., Nemri, A., Harberd, N.P., Jones, J.D. DELLAs control plant immune responses by modulating the balance of jasmonic acid and salicylic acid signaling. *Current Biology*, 2008, *18*(9), 650–655.
175. Choudhury, F.K., Rivero, R.M., Blumwald, E., Mittler, R. Reactive oxygen species, abiotic stress and stress combination. *The Plant Journal*, 2017, *90*(5), 856–867.

CHAPTER 5

Influence of Selective Nanoparticles on the Growth and Yield of Peppermint

FILZA MUSTAFA,[1*] NAYAB NADEEM,[2] and AMNA KOUSAR[3]

[1]*Atta-ur-Rahman School of Applied Biosciences, National University of Sciences and Technology, Islamabad, Pakistan*

[2]*Department of Chemistry, University of Sialkot, Sialkot, Pakistan*

[3]*Department of Biotechnology, University of Sialkot, Pakistan*

*Corresponding author

ABSTRACT

Globally, agricultural sectors are handling environmental threats directly associated with food security. Many crops, including cereals and other minuscule herbaceous plants, aim to surmount these climatic changes. Although different classical and modern methods are available, nanotechnology, especially the use of nanoparticles (NPs), has become an emerging field to study their impact on growth and yield. With distinct physical and chemical characteristics, NPs can elevate plant response toward external stressors. They are excellent carriers of nutrients, both macro and micro, and their combative effect is used against reactive oxygen species. NPs are extensively implemented in the targeted plants, and plants show remarkable results at various growth stages. Peppermint is a well-known medicinal plant facing many hazardous situations in its surrounding environment. Different NPs bring diverse physiological and morphological changes in their bodies, including the growth of roots, shoots, flowering, and seedlings, and even boost the level of phytochemicals and essential oils. This chapter reveals the selective NPs' influence on peppermint's growth and yield.

Peppermint: Novel Strategies for Performance Improvement. Bilal Amhad, Tariq Ahmad Dar, and Khalid Rehman Hakeem (Eds.)

DOI: 10.1201/9781779641557-5

5.1 INTRODUCTION

The word "nano" is derived from the Greek word "Nanos," which means "dwarf" or "little." Nanoparticles (NPs) are particles that are 1–100 nm in size. NPs have been around since the 9th century in Mesopotamia. Artists used NPs to create a gleaming appearance on the surface of pots [1, 2]. Nanotechnology is emerging as a developing field due to its vast range of applications in the fields of optoelectronics, environmental protection, drug delivery, biosensing, and bioimaging. Metal NPs have unique properties (magnetic, optical, catalytic, and electric) which increase the application or demand of NPs in the field of industries. Metal NP manufacturing was projected to be 260,000–309,000 metric tons per year in 2010. Top of the list was alumina, titania, iron, silica, and zinc oxides [3].

The structure and size of NPs depend on the reaction conditions: temperature, pH concentration, time, and reducing agents. By varying reaction conditions, the morphology of NPs changed. Disc, rods, prisms, rings, spherical, bipyramids, platelets, octahedral particles, nanotubes, and triangular prisms [4]. Physical, chemical, and biological processes are used to synthesize silver nanoparticles (Ag NPs). The chemical method involves pulse electrodeposition, chemical coprecipitation, reduction of phytochemicals, and microemulsion. The physical method involves ultrasonication, grinding system, and irradiation. The bottom-up strategy and the top-down approach are both utilized in their synthesis. In the bottom-up approach, NPs are grown from simpler molecules. Sol–gel method, chemical vapor deposition, pyrolysis, molecular condensation, and laser pyrolysis methods are used in this approach. In a top-down approach, NPs are grown from complex molecules. Electro explosions, sputtering, laser ablation, chemical etching, and ball milling methods are used in this approach [5].

Green synthesis follows a bottom-up and biological approach, using plant extracts, algae, microorganisms, and biomass to produce NPs. For NP synthesis, the choice of solvent, the choice of environmentally friendly and reducing agents, and the choice of nontoxic stabilizers and encapsulants are key [5, 6]. NPs have been known for 120 years. NPs have unique properties such as ease of synthesis, high stability, high surface area, nontoxicity, easy tunability and morphology, peak reactivity, size dispersion, electronic and surface properties. NPs have many functions, such as antibacterial, anti-inflammatory, antioxidant, anticancer, antiviral, antifungal, and antiantigen, and are widely used in biologics, drug delivery systems, pharmaceuticals, household healthcare products, and structures [7].

Agricultural production faces an array of threats, encompassing both abiotic and biotic stresses, which are anticipated to have increasingly adverse effects in the future. Consequently, the primary objective for scientists is to enhance growth and productivity in the pursuit of ensuring food security. Furthermore, rising global temperatures are resulting in diminishing agricultural land availability due to inadequate water resources for plant cultivation. This issue is regarded as one of the most significant threats to humanity and all living organisms. The effect of drought on plant growth and yield depends on a variety of factors, including plant species, duration of exposure to stress, current environmental conditions, and the stage of plant development during stress. Drought stress affects plants at the molecular level, causing changes in the expression of stress-sensitive genes, inhibiting growth, and accumulation of organic matter [8].

The main symptom of plant sensitivity to environmental stress is disruption of its photosynthetic system. The existence of stress often results in a decrease in chlorophyll levels and a decrease in the photosynthetic capacity of the leaf due to the degradation of protein and lipid components in the thylakoid membrane. Among the components of the photosystem, photosystem II is the first to respond to environmental stress. In fact, in the photoelectron transport chain, photosystem II is more sensitive to environmental stress than photosystem I. The increase in this phenomenon is due to the location of the water partition in the imaging system II. The intensity of chlorophyll fluorescence is an indicator of the plant's ability to withstand environmental stress, the overall health of the thylakoid membrane, the efficiency and speed of electron transfer from photosystem I to photosystem II, and stress-induced damage to plant growth [9]. In modern agriculture, nanotechnology has been widely used and has played a role in achieving sustainable agriculture. Nanotechnology uses NPs with sizes from 1 to 100 nm (Hatami et al., 2016). NPs can be used to reduce drought in plants, increase nutrients, and improve soil management due to their unique properties, such as small particle size, high surface-to-volume ratio, and the presence of biological or chemical substances. The mint plant (*Mentha piperita* L.) is a member of the Lamiaceae family and occupies an important place in the field of medicinal plants. This herb originates from temperate regions, mainly in Europe, North America, and

North Africa, and has gained worldwide recognition for its cultivation. This plant has different essential oils, from 0.1% to 0.5%. The components of this essential oil are menthol (28–42.2%), menthone (18–28%), menthone (18–24%), menthofuran (19.8%), and methyl acetate (3–10%). In particular, menthol and menthone are the main components of the essential oil of peppermint, known for its good antibacterial properties. Menthol content is the main indicator for assessing the quality of milk essential oil [10].

5.2 INFLUENCE OF NANOPARTICLES ON PLANTS

Nanomaterials have shown a wide range of applications in plant science and improving agricultural capabilities. These multifunctional NPs can be used as smart materials, as herbicides, insecticides, and nanofertilizers to improve plant nutrients and as carriers for the controlled release of agricultural chemicals. In addition, nanomaterials play an important role in accelerating plant adaptation to climate variability and increasing plant stress. In fact, some NPs have unique physical properties that promote plant growth and prevent stress, in addition to their role as simple NPs. The biological effects of NPs depend on their physicochemical properties, the chosen application method (soil, foliar, or hydroponic), and the concentration used. In addition, nanomaterials have shown great potential in the *in vitro* culture of various plant species, including callus induction, organogenesis, somatic embryogenesis, and secondary metabolite production [11].

5.3 NANOPARTICLE UPTAKE AND MOVEMENT WITHIN PLANTS

5.3.1 MODES OF NANOPARTICLE ASSIMILATION BY PLANT SYSTEMS

Within the soil, NPs embark on a series of intricate bio- and geo-transformations that are pivotal in determining their bioavailability and toxicity. Following interactions with plant roots, NPs traverse their way to the aerial parts of plants, ultimately accumulating within cellular or subcellular organelles. The initial phase of this bioaccumulation process involves the adsorption of NPs from the soil by plant roots.

Numerous studies have probed into the mechanisms of plant accumulation, revealing that root adsorption is complemented by subsequent modifications, including crystal phase dissolution, biotransformation, and bioaccumulation. Notably, the size of NPs plays a pivotal role in their absorption, as it dictates their ability to pass through cell wall pores or plant stomata, influencing subsequent transport processes, including plasmodesmata-mediated movement into cells [12].

NP shape also proves to be of significance, with surface area, agglomeration, and reactivity on the cell surface or within plant structures intricately linked to shape. The charge of NPs directly influences their attachment to the plant cell surface, given the typically negative charge of cell walls. Moreover, hydrophobicity on the plant surface significantly impacts the uptake and translocation of NPs. The fundamental structure of nanomaterials further impacts their absorption, translocation, and aggregation within the plant. In light of these considerations, standardizing laboratory experiments becomes imperative to comprehensively evaluate NPs in plant tissues across various dimensions, with a keen focus on their physical–chemical properties [13]. To gain insight into absorption, transfer, and accumulation mechanisms, research into the properties of NPs is essential. Monitoring and tracking NPs is an important step in determining their movement and location in various plant structures and organs [14].

NPs having 3–5 nm diameters have been known to infiltrate plant roots through various mechanisms. Some common methods are the use of capillary forces, osmotic pressure, and direct entry into epidermal cells of plant roots. These epidermal cells of roots, harboring small pores, serve as semi-permeable barriers. These barriers stop large NPs from smooth entry into the plant cellular system. However, in some cases, new pores are induced in the epidermal cell wall, which ultimately supports the entry of some NPs into the cells. After penetrating the cell wall, the NPs begin their apoplastic journey in the extracellular space, eventually reaching the central column, where they can move once through the xylem. To enter the central column, the NPs must pass through the Casparian strip barrier, as shown in Figure 5.1. This transfer occurs through a process that involves the binding of transporter proteins to the endodermal cell membrane and controls endocytosis, pore formation, and transport. The NPs are now incorporated and move from cell to cell *via* plasmodesmata in the cytoplasm. These NPs cannot be included in the Casparian strip.

NPs reach the xylem and are transported to the shoots and returned to the roots *via* the phloem. NPs absorbed by plants are found in epidermal cell walls, cytoplasm, and cortical cell nuclei. The presence of NPs on the root surface can influence nutrient absorption, and the seeds can absorb NPs directly and enter the outer layer through the pore space between the parenchyma cells, and are dissected into the cotyledons [14].

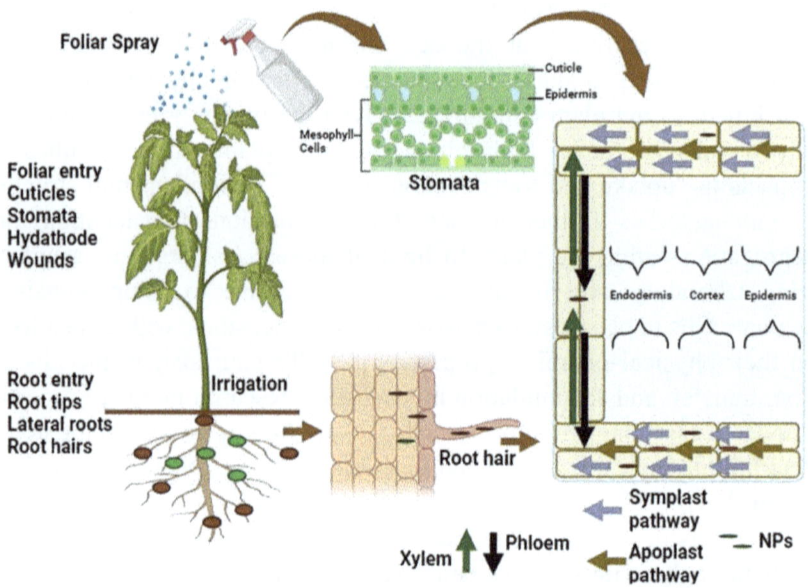

FIGURE 5.1 Illustration depicting the various pathways of nanoparticle uptake and their translocation routes within different parts of plants.

5.3.2 *NANOPARTICLE ENTRY THROUGH PLANT LEAVES*

When NPs are applied to plant leaves, they have the capacity to permeate the leaves through two primary routes: stomata or cuticles. However, the cuticle functions as an initial and formidable leaf barrier, allowing only NPs smaller than <5 nm to pass through. The entry *via* stomata, on the other hand, involves subsequent cellular transport into the plant's vascular system through apoplastic and symplastic routes [15]. The adjacent cells and their cytoplasmic area, known as the symplastic area, predominantly facilitate the translocation of those NPs that range in size from 10 to 50 nm. In contrast, larger NPs, typically measuring between 50 and 200 nm, tend to be transported between cells, following the apoplastic route. Once internalized, phloem sieve tubes are used for interaction between NPs and sugar moieties. This vascular transport *via* the phloem allows NPs to traverse bidirectionally, accumulating to varying degrees in multiple sections of the plant body, like shoot, stems, roots, fruits, grains, and young leaves. These parts function as substantial sinks for the sap [16].

The apoplastic route, often considered the path of least resistance, is nonselective in nature. It has been proven that when many nonessential metal complexes and water-soluble nutrients are translocated, they choose the apoplastic pathway to complete their translocation. Several critical factors, including the method of application, NP size, concentration, and prevailing climate conditions, play pivotal roles in determining the efficient absorption of NPs right after their foliar spraying. Additionally, various aspects have a positive contribution to NPs retention on leaf surface, particularly morphology, presence of trichomes, chemical composition, availability of exudates and waxes on leaf [14].

5.3.4 ENSURING PURITY AND STABILITY OF NPS IN THE PHYSIOLOGICAL ENVIRONMENT

Maintaining the purity and stability of NPs within the physiological environment is of paramount importance. The rate at which NPs accumulate in plant roots can be influenced by a combination of environmental factors and NP properties. Noteworthy studies have demonstrated that when a combination of silver sulfide NPs, potassium chloride, and ammonium thiosulfate was used, it yielded a significant increase in Ag NP concentrations in roots and shoots of *Lactuca sativa,* underscoring the intricate interplay between NPs and the plant environment [17].

5.3.5 UNDERSTANDING THE MECHANISMS OF NANOPARTICLE TRANSLOCATION IN PLANTS

In-depth studies have been undertaken to unravel the delicate pathways governing the absorption, translocation, and accumulation of diverse NPs within various cell organelles and structures of plants. The NPs are translocated and accumulated within plants, and it hinges on several critical factors, including the physiological characteristics and structural intricacies of plant cells, nanomaterials–soil interaction, inherent properties, and NPs stability themselves [18].

The basic function of the plant cell wall is to act as a barrier that regulates NP entry into the cell and influences their solubility and ability to pass through. It is widely acknowledged that the size of cell wall pores serves as a decisive constraint on NP entry into plant cells, with NPs of around 40–50 nm being the size threshold for penetration into the plant body. Another important factor is the nature of NPs that sometimes creates a hindrance during NPs entry into the plant surface, which ultimately unable to find a perfect matching site inside plant cells. So, the attachment of NPs to root and shoot areas of the plant are not possible. For instance, positively charged NPs are strong enough to bind to the cell wall. The rhizosphere and root exudates of plants act as checkpoints and properly scan the morphology and nature of the coating present on the surface of NPs [19]. As NPs accumulate in plant roots, their successful translocation is done to various plant tissues, especially the plant's aboveground parts such as mature seeds. The interplay between plant characteristics and NPs significantly influences the translocation process [20].

5.4 EXPLORING HOW NANOPARTICLES INTERACT WITH PLANTS

5.4.1 IMPACTS OF NANOPARTICLES ON PLANT PHYSIOLOGY

NPs make their entry into the plant system through a variety of avenues, with roots and leaves serving as the primary gateways. Upon penetration, NPs engage with plants at cellular and subcellular levels, inciting a spectrum of changes in the plants' morphological and physiological makeup. These interactions can manifest as either beneficial or adverse, contingent upon the specific characteristics of the targeted host plant and selected class of NPs. Crucially, there are multiple factors to be considered while choosing NPs, especially their size, reactivity, chemical composition, and concentration. Moreover, whether they are found within the plant or on its surface, they wield substantial influence over the impact these NPs will have on plant systems [21].

In order to delve into the ramifications of NPs on plant life, researchers have harnessed various methods for NP application, encompassing soil-based treatments, foliar sprays, and seed treatments. A rich body of evidence has emerged, demonstrating that different NPs have the potential to stimulate seed germination and foster the growth and development of plants, provided they are administered within specific concentration thresholds. These investigations have primarily unfolded in controlled settings, such as growth mediums within laboratory plates, hydroponic systems, or potted conditions. In the pursuit of comprehending how NPs affect plant growth, this discourse will delve into the favorable influence of NPs on seed germination and overall plant growth. Furthermore, we shall explore the constructive effects of NPs in bolstering plant resilience against various stressors [22].

5.4.2 EXPLORING THE MECHANISMS OF INTERACTION BETWEEN NANOPARTICLES AND PLANTS

NPs enter the plant system through various pathways, with roots and leaves being the most prevalent. Upon entry, they engage with plants at cellular and subcellular levels, inducing morphological and physiological alterations. The outcomes of these interactions can be beneficial or detrimental, contingent on the characteristics of the NPs and the specific plant species involved. Crucial factors such as the chemical composition, reactivity, size, and concentration of NPs on or within the plant play a decisive role in determining their overall impact on plant systems [23]. In exploring the impact of NPs on seed germination and plant growth, researchers employed diverse NP application methods, including soil application, foliar spray, and seed treatment. Extensive data indicate that specific NPs, within defined concentration limits, have the potential to improve both seed germination, growth, and development of plants [22].

5.5 INITIATING SEED GERMINATION

Seed germination serves as a fundamental cornerstone for the subsequent growth, development, and overall productivity of plants. Traditional, natural germination methods are undoubtedly effective, yet they invariably demand a considerable amount of time, resulting in less-than-optimal yields. Enter the realm of nanotechnology, where the landscape shifts dramatically, offering a potent means of elevating both germination rates and overall crop yield. A myriad of research endeavors have been undertaken to assess the potential of NPs in augmenting the germination process. Despite these concerted efforts, the precise mechanisms underpinning the way NP treatments bolster seed germination rates continue to elude complete understanding. It is conjectured that NP treatments wield their influence by enhancing seed absorption and water retention, ultimately contributing to the enhancement of seed germination [23].

5.6 UNLOCKING THE POSITIVE EFFECT OF NANOPARTICLES ON PLANT GROWTH AND CROP YIELD

When it comes to crop cultivation, NPs have proven to be very beneficial agents that can produce a wide range of benefits. These include increased rates of seed germination, longer shoots and roots, increased fruit yields, higher metabolite concentrations, and notable increases in seedling and vegetative biomass in a variety of crops. Furthermore, NPs have permanently altered a number of biochemical variables connected to the growth and development of plants. They have been crucial in increasing the rate of photosynthetic activity and improving the efficiency of nitrogen usage in a number of important crop varieties, such as soybeans, spinach, and peanuts [24].

Beyond these facets, NPs have become known for their pivotal role in ameliorating nutrient uptake while simultaneously fortifying plants against an array of diseases and the rigors of abiotic stressors. These minuscule powerhouses are adept at influencing plant growth and development by orchestrating alterations in various physiological processes inherent to plants. It is worth noting that while NPs indeed yield a treasure trove of benefits, their impact can shift into the realm of toxicity when concentrations surpass a certain threshold. Plant toxicity studies have cast a discerning eye on the repercussions, scrutinizing their effects on germination rates and biomass accumulation, ultimately revealing that NPs have the potential to exert substantial effects on plants [11].

5.7 HARNESSING THE BENEFICIAL EFFECTS OF NANOPARTICLES ON PHOTOSYNTHESIS

Photosynthesis, that indispensable process converting radiant energy into chemical energy, stands as the lifeblood of Earth's plants. It serves as a direct energy source for some and indirectly fuels the energy needs of others, as it underpins the foundation of food chains. Figure 5.2 illustrates the general mechanism by which, within the chloroplasts, this remarkable feat is accomplished, with chlorophyll, H_2O, and carbon dioxide (CO_2) serving as the raw materials for the conversion of light energy into chemical bonds within sugar molecules. Enter the realm of NPs, and a fascinating narrative unfolds. An array of studies has illuminated the extraordinary capacity of metal NPs, when applied topically to foliage, to significantly augment chlorophyll content in plants. This augmentation equips plants with an enhanced ability to synthesize complex compounds, facilitating superior light absorption and ultimately elevating the titanium dioxide (TiO_2) NP, which is well-known for its photocatalytic capabilities and capacity to initiate oxidation–reduction reactions, is spearheading this research. For their part in mediating charge transfer between TiO_2 NPs and light-harvesting complexes II, TiO_2 NPs have drawn attention. TiO_2 NPs have been observed to have a noticeable impact on the photosynthetic efficiency of spinach, providing an example of the application. More light absorption, faster energy transfer, and more intense energy transformation are the results. Surprisingly, adding TiO_2 NPs can also prevent chloroplast aging, extending the time that photosynthesis functions properly. Despite both visible and ultraviolet (UV) light, spinach's electron transport chain, oxygen-evolving and photophosphorylation activities, and photoreduction function of chlorophyll have all been improved by nanoanatase TiO_2 [27] efficiency of photosynthesis [25, 26].

In addition to TiO_2 NPs, zinc oxide (ZnO) NPs have also left an indelible mark. A notable study observed a substantial upsurge in soluble protein and chlorophyll content in ZnO NP-treated plants, marking an increase of 25% and 34.5%, respectively, when compared to control groups. Furthermore, through an aerosol-foliar spray method, chlorophyll content soared, ranging from a 62.67% to an astounding 227.42% increase with escalating concentrations of TiO_2 NPs up to 500 mg kg^{-1}. When TiO_2 NPs were introduced into soil, they demonstrated their potential, yielding a remarkable 216.29% increase in chlorophyll content at a concentration of 750 mg kg^{-1}[29] [27].

5.8 ENHANCING PLANT STRESS TOLERANCE WITH NANOPARTICLES

Plants are remarkably adaptive and resilient to harsh environments, such as heat stress, freezing, drought, and salt. Numerous studies have been

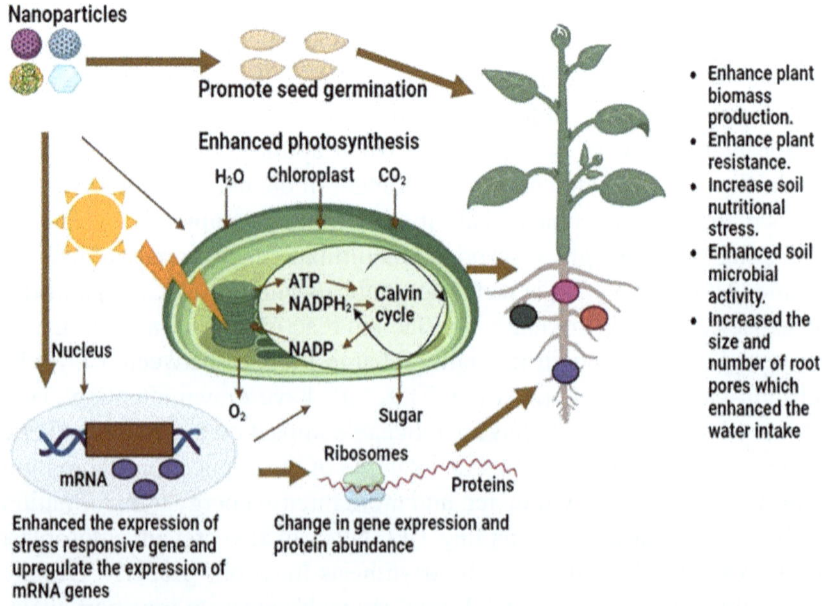

FIGURE 5.2 Positive effects of nanoparticles on the development and growth of plants. The proper concentration of nanoparticles alters a number of physiological functions, improving photosynthesis and seed germination. Furthermore, the expression of genes and miRNAs is influenced by nanoparticles, which enhances plant biomass and resilience to stress.

conducted, exploring both the cellular and molecular levels of plant resistance to abiotic stresses. Plants respond to abiotic stress by temporarily elevating their levels of cytoplasmic calcium, producing abscisic acid, generating reactive oxygen species (ROS), and activating pathways linked to mitogen-activated protein kinase. Plants use complex mechanisms to cope with increasing stress, such as regulating gene expression specific to stress and controlling proteins that are essential for cellular defense against injury (see Figure 5.3) [7].

Increased crop yields in difficult environments can be achieved with nanofertilizers, as demonstrated by recent advancements in nanomaterial engineering. Crop productivity is severely limited in about 23% of the world's cultivated lands due to severe salinity stress. In tomatoes under NaCl-induced stress, silica NPs, for example, have shown the capacity to improve proline accumulation and chlorophyll content, as well as to increase fresh and dry plant weight. The application of iron sulfate NPs topically has also shown encouraging results in improving sunflower cultivars' ability to withstand salinity stress. This application increases the amount of leaf area, increases the maximum photochemical efficiency of photosystem II (Fv/Fm), increases the net assimilation rate of CO_2, increases the content of chlorophyll, increases shoot dry weight, increases substomatal CO_2 concentration, and increases iron (Fe). It also depletes the leaves of sodium (Na) [7, 28].

In onion seedlings, TiO_2 NPs have been found to stimulate the expression of antioxidant enzymes. When TiO_2 NPs and Ag NPs were used together, the activity of these enzymes was increased even more. TiO_2 concentrations between 10 and 30 μg/mL were linked to enhanced onion seedling growth and seed germination, whereas concentrations between 40 and 50 μg/mL showed suppressive effects. The investigation also discovered that at lower TiO_2 concentrations, enhanced superoxide dismutase (SOD) activity was induced, along with hydrolysis enzyme (amylase), catalase, and peroxidase (POD) activity. Furthermore, the foliar application of TiO_2

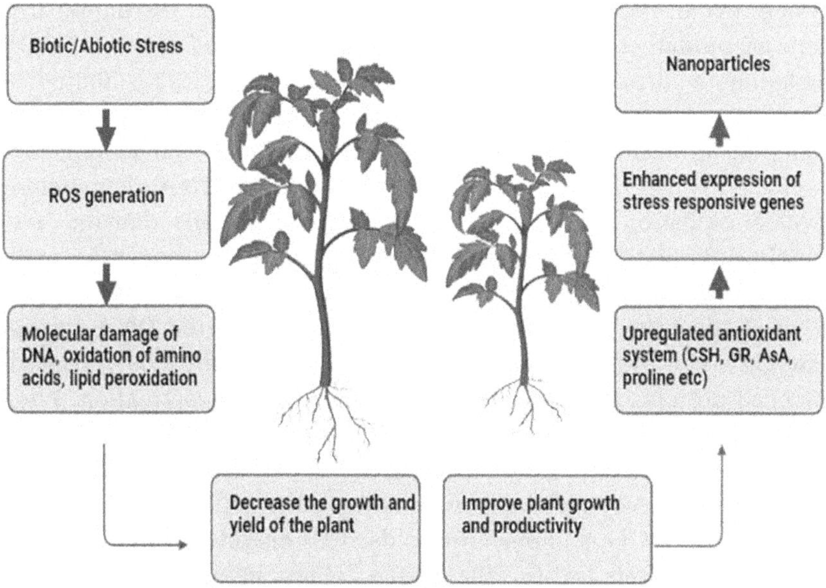

FIGURE 5.3 Nanoparticles enhance plants' ability to mitigate stress in diverse environmental conditions through general mechanisms.

NPs on wheat has shown promising results in mitigating the negative consequences of drought stress. Plant height, ear number, ear weight, seed count, final yield, biomass, gluten-included harvest index, and starch content under drought stress conditions have all improved as a result of this application [23].

The effects of drought stress may be mitigated by using silicon, it has been discovered. According to earlier research, wheat can be protected against the negative effects of drought stress by a small dose of sodium silicate (1.0 mM). It is known that silicon can improve leaf water retention, raise the content of chlorophyll in leaves, and partially boost shoot growth in stressed plants, though the precise mechanism is still unknown. It also helps to lower wheat's membrane lipid peroxidation [7].

5.9 EFFECTS OF NANOPARTICLES ON PLANT HEALTH AND THEIR INDUCED SURVIVAL MECHANISMS

NPs infiltrate plant cells and interfere with the electron transport system cycle in chloroplasts and mitochondria. This disruption triggers an oxidative burst characterized by heightened levels of ROS, including hydroxyl radicals, superoxide-free radicals, nonradical singlet oxygen, and hydrogen peroxide. While ROS often serve as signaling molecules, their excessive presence can lead to oxidative stress. This heightened ROS presence induces DNA degradation, protein oxidation, lipid peroxidation, and membrane damage, ultimately culminating in programmed cell death. Consequently, metal and metal-based NPs have the potential to induce oxidative stress in a broad spectrum of plant species. The ROS-mediated DNA damage extends to chloroplast and mitochondrial DNA, leading to alterations in encoded proteins and subsequent malfunction and deactivation. When *Ricinus communis* seeds are subjected to treatment with AgNPs, there is an elevation in ROS production along with an active engagement of antioxidant defense mechanisms, involving the activation of enzymes such as SOD, POD, and phenolic acids. This heightened production of phenolic chemicals fortifies the plant's defense against infections and, interestingly, prompts an increase in root length, albeit up to a specific threshold [29].

5.10 UTILIZING NANOPARTICLES FOR ADVANCING SUSTAINABLE AGRICULTURE

The incorporation of NPs in sustainable agriculture holds the promise of revolutionizing the agricultural and food sectors through innovative approaches to disease management, disease detection, and enhancement of plant nutrient absorption. This groundbreaking strategy seeks to enhance agricultural efficiency while reducing reliance on conventional inputs such as fertilizers, insecticides, and herbicides. Often referred to as "magic bullets," NPs are engineered to specifically target certain cellular organelles within plants, ensuring the efficient delivery of their intended effects. Exhibiting characteristics akin to nanopesticides, nanofertilizers, and herbicides, these NPs are crafted to address plant diseases and stimulate robust plant development [30].

5.10.1 NANOFERTILIZERS

Nanofertilizers, in particular, play a crucial role in elevating crop yields. Their impact is reflected in the content and reactivity of phytohormones, serving as key indicators of plant physiology. For example, the application of CeO_2 NPs improves biochemical factors such as amino acids, nonreducing sugars, fatty acids, and phenolic complexes in diverse plant species. Nanofertilizers optimize nutrient productivity by delivering nutrients in alignment with the specific requirements of crops. Several research studies have illustrated the superior effectiveness of nanofertilizers in comparison to traditional chemical fertilizers [31].

5.10.2 NANOHERBICIDES

In the realm of sustainable agriculture, nanoherbicides also emerge as valuable tools for weed control and management. Remarkably, even minute quantities of NPs can bind with soil and effectively eliminate weeds from the environment, all while avoiding toxicity concerns and the unintended growth of undesirable plant species. The innovative utilization of NPs presents a hopeful path toward encouraging sustainable and eco-friendly agricultural methods [32].

5.11 ROLE OF SELECTIVE NANOPARTICLES ON PEPPERMINT GROWTH AND YIELD

The agricultural sector is confronted with numerous challenges due to the decline in the production of food and the degradation of agricultural land. There are different techniques and tools that can be implemented on various crops to improve their development, growth, and yield. Similar to other staple crops, various small herbaceous plants are also subjected to stress or environmental elements that affect their development. *M. piperita* is a crucial and widely utilized medicinal plant that is traditionally utilized for medicinal purposes. It is used for a variety of health and food reasons because it is rich in phytochemicals of various kinds as well as essential oils. A variety of effective methods can be employed to boost the productivity and growth of peppermint. One method is the use of NPs (Figure 5.4) [9].

NPs are used in combination to enhance the growth of different parts of peppermint, like leaves, roots, shoots, flowers, and seeds. Generally, these NPs are responsible for improving a plant's ability to withstand stresses both externally and internally [33]. NPs are instilled into the plant's roots to enhance the intake of nutrients. Others are incorporated in the forms of foliar sprays that improve transpiration, photosynthesis, and gaseous exchange in branches and leaves. There is a myriad of NPs that are mixed

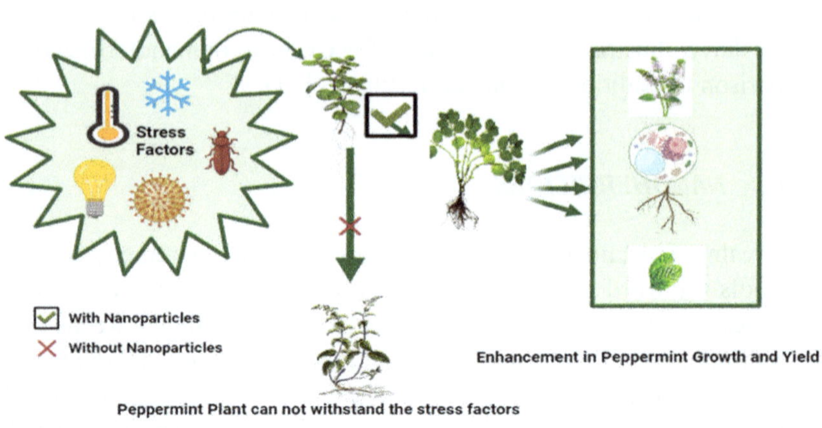

FIGURE 5.4 In the presence of environmental stress factors, nanoparticles elevate plant defense mechanisms and bring positive changes in the growth and yield of peppermint.

with other fertilizers or with certain microorganisms, such as bacteria and fungi. Scientifically, the positive effect of NPs on the vegetative growth as well as the yield of peppermint has been proven. They further highlight that the amount of phytochemical compositions is more pronounced in plants that are treated with NPs compared to ones that are not treated. Some typically used NPs to regulate the developmental stages of peppermint under stress conditions are thoroughly explained here [9].

5.11.1 ROLE OF IRON NANOPARTICLES IN PEPPERMINT GROWTH

NPs are increasingly becoming demanded agricultural crop fertilizers due to their high absorption, their nanostructure, as well as their capacity to reduce nutrient deficiencies in plants. NPs made of different metals are used in medicine, industry, and agriculture. One of these is the iron oxide (Fe_2O_3) NPs popular for their huge surface area and a highly reactive nature [34]. *M. piperita*, also known as peppermint, is affected by changes in the environment, particularly due to salinity. The high level of salinity caused many issues like cell membrane degradation, lower photosynthesis, less water absorption, and a decrease in leaf formation [9]. Iron is an essential component needed for photosynthesis. Approximately 20% of the cells responsible for photosynthesis are rich in iron. When plants are cultivated under saline conditions, the iron content drops below the recommended levels. In the end, all of the related mechanisms revealed a disturbance. When iron NPs are employed, plant cells absorb them and ultimately increase the level of iron in the leaves of peppermint. When the iron content has reached its equilibrium, the synthesis of cytochrome and compounds containing heme, as well as chlorophyll, and cluster formation are initiated. It is directly responsible for enhancing the productivity and growth of peppermint [9].

Researchers utilized iron NPs in two ways. One was foliar spraying, and the other was the incorporation into the soil. Both methods produced remarkable results. They found that when peppermint was grown under salinity conditions, the sodium and phosphorus levels were increased to some degree, while the gradual increase in salt stress reduced the zinc, iron, potassium, and calcium content. The salt stress was associated with the formation of reactive oxygen compounds, which caused further damage to the plasma membrane as well as the integrity of the membrane [9].

In the wake of this injury caused by this damage, iron was sucked out of cells. Na^+ ions began to build up in the body parts of plants, like leaves and shoots. Additionally, potassium ions are eliminated from plants because the plant cannot store potassium in its shoots. Therefore, iron NPs treatment in the form of foliar sprays increased the potassium and iron levels, as well as zinc and calcium content in peppermint under salinity. The researchers concluded that the use of NPs, particularly Fe_2O_3 NPs aids the plants to absorb a high amount of micronutrients, which plants were incapable of doing in salinity. Furthermore, the use of iron NPs reduced CAT, SOD, and GPOX formation [35].

Another scientific team evaluated the effects of iron NPs on the growth of *M. piperita* under salinity. They revealed the negative impact of salts, including a reduction in gas exchange rate and a smaller surface area for leaves, closure of stomatal pores, and the growth of roots [36]. Additionally, they assessed the effects of stress on various areas of the body, focusing particularly on its stems, flowers, and roots, as well as on the leaves [37]. Abiotic Stress, specifically drought, hinders the yield and growth of the peppermint plant. The gradual increase in drought is correlated with reduced development and growth of the roots as well as leaves from peppermint. This is due to the less gaseous exchange and transpiration because of water content is lower than the average. Thus, research was conducted to address drought stress and increase the resistance of plants with the aid of growth stimulators, specifically kitoplaus, which was enriched with chitosen. Chitosen decreased the stress caused by drought and also enhanced the ability of gaseous exchange and increased the efficiency of transpiration. During this experiment, iron was employed in the form coating on chitosen [35].

They analyzed the effects of Kitoplaus, which is a growth stimulator, and chitosen-coated Fe_2O_3 NPs on peppermint's growth, as well as yield and development. Their wet lab experiments were based on different concentrations of growth stimulants, Fe_2O_3 NPs. In this regard, they employed peppermint rhizomes and sprayed the targeted NPs at specific intervals. After 15 days, they determined various aspects, including the length of the stem, diameter, and number of nodes, distance between internodes, the weight of the flower cluster, dry and fresh weight of aerial parts, etc. So, the researchers concluded that the stress of drought in peppermint can be reduced through the spraying of Kitoplaus and Fe_2O_3 NPs on leaves since they both increased the rate of photosynthesis as well as transpiration through leaves of peppermint. Therefore, these Fe_2O_3 NPs could be an alternative strategy in the modern agricultural industries to shield plants from stress caused by drought and could be utilized to achieve the desired yield in adverse conditions [35].

The yield of peppermint is an additional aspect that decreased because of stress, particularly due to salinity. When the salinity of the environment is high, the oil composition within *M. piperita* is less than the normal levels of essential oils that are produced in its aerial components [35].

5.11.2 ROLE OF MAGNETITE NANOPARTICLES IN PEPPERMINT GROWTH

Magnetite nanoparticles (MNPs) are highly beneficial for regulating the growth stages of peppermint. They employed MNPs that are highly reactive to boost iron deficiency in peppermint. They discovered that the deficiency of iron found in peppermint was directly linked with the inhibition of various photosynthetic pathways. Due to the absence of photosynthesis, the plant growth is stunted and then produces less quantity of phytochemicals and essential oils. In this study, they discovered that the use of MNPs accelerated every stage of growth in peppermint, particularly a rise in the fresh and dry weight of the leaves was noticed [37].

The second aspect was the determination of rates of photosynthesis, which were higher in plants that were treated with MNPs than those of untreated or control plants. The sprayed 15ppm MNPs on peppermint, and the iron levels increased to 5.6 from 2.8 and as high as 5.9 from 3.2 per milliliter of plants that were not treated. This is not all, sprays of MNPs carried out on plants increased the concentrations of zinc, manganese, and copper in the peppermint [33, 37].

5.11.3 ROLE OF TITANIUM DIOXIDE NANOPARTICLES IN PEPPERMINT GROWTH AND YIELD

NPs are a great addition to peppermint. They could also be used in mix with other stress-reducing compounds like arbuscular mycorrhizal fungi to increase the production and growth. This study has demonstrated the negative impact of salinity on decreasing the various natural photosynthesis pathways in peppermint that are associated with the formation of reactive oxygen compounds. Additionally, they discussed how salinity poses an environmental danger for peppermint, which caused the closing of stomata due to stomatal damage. It also reduces the absorption of water and increases the speed of photorespiration [38].

In this study, arbuscular mycorrhizal fungi were utilized as a biofertilizer in order to enhance the plant's defense mechanisms against extreme conditions. NPs were utilized due to their tiny dimensions, their speed of delivery, and catalytic abilities to enhance the productivity of plants and increase their growth during salinity. NPs of TiO_2 are known for their stress-modulating properties, which help the plants to increase the growth of their roots. It also ensured that the nutrient intake is greater in peppermint with the help of TiO_2 NPs. Arbuscular mycorrhizal fungi, as well as TiO_2 NPs work as stimulants for peppermint. In addition, the plant has shown a significant improvement in the growth of its roots and has adapted to nutritional deficiencies too [38].

Other advantages that TiO_2 NPs bring to peppermint include controlling the antioxidant enzyme's actions in the transformation of inorganic nitrogen into organic nitrogen in the form of proteins, enhance the biosynthesis of chlorophyll, and improve the cell membrane stability and the assimilation of carbon from photosynthetic sources. Recent studies have confirmed that titanium NPs made of TiO_2 are being studied for their ability to enhance the productivity of the growth process and enhance biochemical levels in medicinal plants, such as peppermint [38].

In peppermint, the TiO_2 NPs readily got into plant cells. They gradually boosted the growth rate by increasing the dry weight and yield, and also generating different phytochemical compounds in peppermint. The plant also benefited from the ability to grow taller by regulating the growth of the stem in conjunction with root development and growth together with Myco-root + TiO_2 NPs [38].

Another research work was performed to examine the impact of TiO_2 NPs on peppermint's growth and the parameters for how they affect biochemical properties of peppermint. In this study, scientists have identified the crucial role played by TiO_2 in controlling and increasing the growth of peppermint. First, they elucidated the role of TiO_2 NPs in seed germination. In this study, they inoculated peppermint seedlings with TiO_2 on petri dishes. They also observed that after a period of eight days, the rate of seed germination was significantly greater than those that were not inoculated. They also reported that these NPs helped to increase the rate of germination in seeds of peppermint, too. The next step was to determine the growth of peppermint shoots. They treated the peppermint's shoots with two different levels of TiO_2 and observed an increase of 21% in the length of the shoot. Additionally, they examine the effect that TiO_2 NPs within those peppermint plants in which the growth is reduced due to methyl jasmonate. As methyl jasmonate is associated with the reduced growth of peppermint shoots [39].

It was also noted that TiO_2 NPs were good enough to increase dry and fresh weight. The combined effect of TiO_2 together with hormones such as methyl jasmonate, as well as salicylic acid, was also examined and, surprisingly, it produced amazing results. The NPs are responsible for helping aid in the production of these growth hormones, particularly in the absence of their ability to boost the growth rate and development. These combinations ultimately were linked with the synthesis of secondary metabolites of peppermint, too. The introduction of TiO_2 in the plant increased its capacity to absorb nutrients [39].

5.11.4 ROLE OF AU–AG NANOPARTICLES IN PEPPERMINT GROWTH AND YIELD

NPs come from many sources, such as microorganisms, metals, and plants; however, metal NPs pose an extremely dangerous threat to the ecosystem. They are because they undergo low degradation rates and are poisonous in the natural world. Chemical NPs also contain various reagents, and their removal and purification are difficult to manage. NPs derived from plant extracts or microorganisms are environmentally friendly and cost-effective. Au–Ag NPs can be prepared in both methods: biological and chemical. Chemical Au–Ag NPs pose a threat to the growth of plants production, development, and yield; however, Au–Ag NPs made from plants can be beneficial to plant health and increase yield and growth of plants [40].

Salinity is a highly disruptive aspect of cell division and development in plants, especially peppermint. The high level of chloride and sodium ions is directly connected with the decrease of the other ions essential to life, including ammonium, calcium, nitrate, and potassium. The concentration of these ions becomes automatically lower when there are sodium and chloride ions present in peppermint. Additionally, salinity can disrupt enzyme actions and the structure of membranes. This causes the plant to lose different processes, like metabolism. The rate of damage to photosynthesis is much lower than normal, and the reduction in height of plants is measurable. This indicates that all physiological and biochemical processes are not functioning effectively inside the body of peppermint, hence the area of leaves and the number of leaves are lower [40].

There are a variety of growth hormones, such as auxin, gibberellin, and cytokinin. Auxin is added externally to help boost development and growth. In comparison to the most recent NPs released, this release is slow. Au–Ag NPs are utilized in lower concentrations and enhance the plant's dry and fresh weights in salinity conditions. Ag NPs in this alloy can be responsible for increasing the weight of plants, their length, as well as height and stem growth. Additionally, the silver particles are able to lessen stress from drowning, particularly the loss of weight and the reduction in altitude [40].

The physiological properties of peppermint were enhanced, and the quantity of chlorophyll, which was high in peppermint upon Ag NPs inoculation. The Au–Ag NPs in the aggregate influenced different internal and external growth factors in peppermint. They are responsible for enhancing enzyme activity, the rate of photosynthesis, and the creation of antioxidants. In the end, these Au–Ag NPs are able to decrease the effect of salinity on growth, and increase the resistance to extreme stress conditions [40].

5.11.5 ROLE OF MAGNESIUM OXIDE NANOPARTICLES IN PEPPERMINT GROWTH AND YIELD

One of the main threats to the growth of peppermint is heavy metal stress, especially cadmium stress, because it is directly linked to a reduction in growth and a decrease in physical performance and physiological efficiency of plants. Additionally, cadmium also damages the shoot and shoot tissues in the plant and affects the normal development and growth phases of peppermint leaves. It also comes from the development of many toxic enzymatic organisms that eventually slow the photosynthetic rate as well as degrade the chlorophyll pigment. Another alarming issue is the creation of ROS that can be responsible for reducing the expansion of leaves, their less growth rate, and chlorophyll concentration. If peppermint plants are grown in soils that are high in cadmium stress, magnesium plays an important role in enhancing the plant's defense mechanism against environmental stresses. The research conducted was to study the effects of the application of magnesium foliar to peppermint on growth production, development, yield, and biological processes [41].

In this study, first, they cultivated peppermint and found that different internal mechanisms were affected, specifically the chlorophyll content of leaves, which is a major factor in essential oil concentration, and so on. In order to enhance the defense mechanism of peppermint, the spraying of magnesium NPs carried out that boosted the leaves' surface area, dry and fresh biomass development of different parts, as well as increased the concentration of essential oils beneficial to the plant. Magnesium oxide (MgO) NPs were effective in increasing plant resistance against cadmium stress [42]. Furthermore, MgO NPs are capable of modulating oxidative stress. They also enhance the osmotic regulation, which increases magnesium levels and reduces the concentration of cadmium in the plant tissues [41].

5.11.6 ROLE OF COPPER NANOPARTICLES IN PEPPERMINT GROWTH AND YIELD

Peppermint is well-known for its therapeutic properties and multiple health benefits. Numerous environmental factors cause a decrease in the development and growth of peppermint. Therefore, the use of NPs on specific plants is a simple and affordable process. Along with other NPs, copper NPs are an emerging class of NPs that can alter the development, growth, and yield. Scientists carried out a study of research on how copper NPs work and examined their impact on the growing and developing of the vegetative components of peppermint. In their research, they found that the use of copper NPs to peppermint dramatically altered various traits such as branch and leaf number, leaf size, as well as the amount of essential oils and phytochemicals [43, 44].

The study revealed that all plants treated with copper NPs demonstrated greater growth. Also, it was observed that the levels of pigments that are photosynthetic were increased to some extent due to the fact that the vegetative parts became healthier and grew properly. In addition, they explain that copper NPs should be used in limited quantities, not in excessive amounts, since greater concentrations of copper cause lower plant growth. Another key parameter is the use of copper NPs in conjugation with fertilizers, such as Algacifo fertilizer, used to monitor the behavior of NPs in relation to peppermint. Overall, a positive view of copper NPs was observed in comparison to peppermint [44].

5.12 CONCLUSION

NPs not only improve the internal defense mechanisms of plants against environmental stress but also improve physical activity, growth, and productivity. However, the timing and method of NPs application are important to plant health. Some are applied by foliar spray, while others are injected into plant roots. All of these factors are closely related to the increase in gas exchange, the ease of absorption of nutrients, and the speed of metabolism. *M. piperita* is a delicate plant with small roots and leaves. Excessive use of chemicals reduces the content of phytochemicals and essential oils in its parts. So, selective NPs, including Fe_2O_3, MgO, ZnO, and TiO_2, help plant internal mechanisms. These NPs make their way inside the plant body and efficiently perform their activities, due to which plants show an improvement in growth and developmental stages.

KEYWORDS

- **nanotechnology**
- **nanoparticles**
- **titanium dioxide**
- **iron oxide nanoparticles**
- **zinc oxide**
- **magnetite nanoparticles**
- **magnesium oxide**
- **plant growth hormones**
- **heavy metal**
- **salinity**
- **mycorrhizal fungi**

REFERENCES

1. McNeil, S.E. Nanotechnology for the biologist. *Journal of Leukocyte Biology*, 2005, *78*(3), 585–594.
2. Bilal, M., Ihsanullah, I., Shah, M.U.H., Reddy, A.V.B., Aminabhavi, T.M. Recent advances in the removal of dyes from wastewater using low-cost adsorbents. *Journal of Environmental Management*, 2022, *321*, 115981.
3. Kolhatkar, A.G., Jamison, A.C., Litvinov, D., Willson, R.C., Lee, T.R. Tuning the magnetic properties of nanoparticles. *International Journal of Molecular Sciences*, 2013, *14*(8), 15977–16009.
4. Hajipour, M.J., Fromm, K.M., Ashkarran, A.A., de Aberasturi, D.J., de Larramendi, I.R., Rojo, T., Serpooshan, V., Parak, W.J., Mahmoudi, M. Antibacterial properties of nanoparticles. *Trends in Biotechnology*, 2012, *30*(10), 499–511.
5. Rane, A.V., Kanny, K., Abitha, V., Thomas, S. Methods for synthesis of nanoparticles and fabrication of nanocomposites. In: *Synthesis of Inorganic Nanomaterials*. Elsevier. 2018; pp. 121–139.
6. De Carvalho, J., De Medeiros, S., Morales, M., Dantas, A., Carriço, A. Synthesis of magnetite nanoparticles by high energy ball milling. *Applied Surface Science*, 2013, *275*, 84–87.
7. Iqbal, M.S., Singh, A.K., Singh, S.P., Ansari, M.I. Nanoparticles and plant interaction with respect to stress response. *Nanomaterials and Environmental Biotechnology*, 2020, 1–15.
8. Pestovsky, Y.S., Martínez-Antonio, A. The use of nanoparticles and nanoformulations in agriculture. *Journal of Nanoscience and Nanotechnology*, 2017, *17*(12), 8699–8730.
9. Askary, M., Talebi, S.M., Amini, F., Bangan, A.D.B. Effects of iron nanoparticles on Mentha piperita L. under salinity stress. *Biologija*, 2017, *63*(1). https://doi.org/10.6001/BIOLOGIJA.V63I1.3476
10. Amin, A.-R.H., Bayoumi, A.E.-D., Dimetry, N.Z., Youssef, D.A. Efficiency of Nano-formulations of neem and peppermint oils on the bionomics and enzymatic activities of Agrotis ipsilon larvae (Lepidoptera: Noctuidae). *Natural Resource Ecology and Management*, 2019, *4*, 102–111.
11. Zhou, P., Adeel, M., Shakoor, N., Guo, M., Hao, Y., Azeem, I., Li, M., Liu, M., Rui, Y. Application of nanoparticles alleviates heavy metals stress and promotes plant growth: An overview. *Nanomaterials*, 2020, *11*(1), 26.
12. Jackson, B.P., Bugge, D., Ranville, J.F., Chen, C.Y. Bioavailability, toxicity, and bioaccumulation of quantum dot nanoparticles to the amphipod Leptocheirus plumulosus. *Environmental Science & Technology*, 2012, *46*(10), 5550–5556.
13. Mun, T., Koo, J.Y., Lee, J., Kim, S.J., Umarji, G., Amalnerkar, D., Lee, W. Resistive-type lanthanum ferrite oxygen sensor based on nanoparticle-assimilated nanofiber architecture. *Sensors and Actuators B: Chemical*, 2020, *324*, 128712.
14. Chokkareddy, R., Redhi, G.G. Green synthesis of metal nanoparticles and its reaction mechanisms. In: *Green Metal Nanoparticles: Synthesis, Characterization and Their Applications*. 2018; pp. 113–139.

15. Khan, M.R., Adam, V., Rizvi, T.F., Zhang, B., Ahamad, F., Jośko, I., Zhu, Y., Yang, M., Mao, C. Nanoparticle–plant interactions: Two-way traffic. *Small*, 2019, *15*(37), 1901794.
16. Mishra, V., Mishra, R.K., Dikshit, A., Pandey, A.C. Interactions of nanoparticles with plants: An emerging prospective in the agriculture industry. In: *Emerging Technologies and Management of Crop Stress Tolerance*. Elsevier. 2014; pp. 159–180.
17. Lin, P.-C., Lin, S., Wang, P.C., Sridhar, R. Techniques for physicochemical characterization of nanomaterials. *Biotechnology Advances*, 2014, *32*(4), 711–726.
18. Hubbard, J.D., Lui, A., Landry, M.P. Multiscale and multidisciplinary approach to understanding nanoparticle transport in plants. *Current Opinion in Chemical Engineering*, 2020, *30*, 135–143.
19. Raliya, R., Franke, C., Chavalmane, S., Nair, R., Reed, N., Biswas, P. Quantitative understanding of nanoparticle uptake in watermelon plants. *Frontiers in Plant Science*, 2016, *7*, 1288.
20. Kumar, S., Masurkar, P., Sravani, B., Bag, D., Sharma, K.R., Singh, P., Korra, T., Meena, M., Swapnil, P., Rajput, V.D. A review on phytotoxicity and defense mechanism of silver nanoparticles (AgNPs) on plants. *Journal of Nanoparticle Research*, 2023, *25*(4), 54.
21. Sami, F., Siddiqui, H., Hayat, S. Impact of silver nanoparticles on plant physiology: A critical review. In: *Sustainable Agriculture Reviews 41: Nanotechnology for Plant Growth and Development*. 2020; pp. 111–127.
22. Ahmad, A., Hashmi, S.S., Palma, J.M., Corpas, F.J. Influence of metallic, metallic oxide, and organic nanoparticles on plant physiology. *Chemosphere*, 2022, *290*, 133329.
23. Landa, P. Positive effects of metallic nanoparticles on plants: Overview of involved mechanisms. *Plant Physiology and Biochemistry*, 2021, *161*, 12–24.
24. Goswami, P., Mathur, J., Srivastava, N. Silica nanoparticles as novel sustainable approach for plant growth and crop protection. *Heliyon*, 2022, *8*(7), e09908.
25. Wang, X., Yang, X., Chen, S., Li, Q., Wang, W., Hou, C., Gao, X., Wang, L., Wang, S. Zinc oxide nanoparticles affect biomass accumulation and photosynthesis in Arabidopsis. *Frontiers in Plant Science*, 2016, *6*, 1243.
26. Lin, D., Xing, B. Phytotoxicity of nanoparticles: Inhibition of seed germination and root growth. *Environmental Pollution*, 2007, *150*(2), 243–250.
27. Sabir, S., Arshad, M., Chaudhari, S.K. Zinc oxide nanoparticles for revolutionizing agriculture: Synthesis and applications. *The Scientific World Journal*, 2014, *2014*, 925494.
28. Jalil, S.U., Ansari, M.I. Nanoparticles and abiotic stress tolerance in plants: Synthesis, action, and signaling mechanisms. *Plant Signaling Molecules*, 2019, 549–561.
29. Li, Y., Zhang, P., Li, M., Shakoor, N., Adeel, M., Zhou, P., Guo, M., Jiang, Y., Zhao, W., Lou, B. Application and mechanisms of metal-based nanoparticles in the control of bacterial and fungal crop diseases. *Pest Management Science*, 2023, *79*(1), 21–36.
30. Hazarika, A., Yadav, M., Yadav, D.K., Yadav, H.S. An overview of the role of nanoparticles in sustainable agriculture. *Biocatalysis and Agricultural Biotechnology*, 2022, *43*, 102399.
31. Xin, X., Judy, J.D., Sumerlin, B.B., He, Z. Nano-enabled agriculture: From nanoparticles to smart nanodelivery systems. *Environmental Chemistry*, 2020, *17*(6), 413–425.

32. Goswami, P., Mathur, J. Positive and negative effects of nanoparticles on plants and their applications in agriculture. *Plant Science Today*, 2019, *6*(2), 232–242.
33. Mohammadipour, R., Sedaghat Hoor, S., Mahboub-Khomami, A. Effect of application of iron fertilizers in two methods foliar and soil applications on growth characteristics of Spathiphyllum illusion. *European Journal of Experimental Biology*, 2013, *3*, 232–240.
34. Wannoussa, W., Masy, T., Lambert, S., Heinrichs, B., Tasseroul, L., Al-Ahmad, A.E., Weekers, F., Thonart, P., Hiligsmann, S. Effect of iron nanoparticles synthesized by a Sol-Gel process on Rhodococcus erythropolis T902. 1 for biphenyl degradation. *Journal of Water Resource and Protection*, 2015, *7*, 264–277.
35. Torabi Giglou, M., Heydarnajad Giglou, R., Azarmi, R., Salimi, G., Maleki Lajayer, H., Mokhtari, A.M., Bagherian, M. Effects of Kitoplus® and Chitosan coated Iron Nano-oxide on morpho-physiological properties of peppermint under drought stress. *Journal of Vegetables Sciences*, 2023, *6*(2), 135–146.
36. Mahto, R., Rani, P., Bhardwaj, R., Singh, R.K., Prasad, S.K., Rakshit, A. Nanotechnology: A potential approach for abiotic stress management. In: *Advances in Nano-Fertilizers and Nano-Pesticides in Agriculture*. Elsevier. 2021; pp. 249–264.
37. El-Kereti, M., El-feky, S.A., Khater, M.S., Osman, Y., El-Sherbini, E.-S. Effect of He Ne laser irradiation and Fe3O4 NPs foliar spray on growth and yield of sweet basil plant. *International Journal of Indigenous Medicinal Plants*, 2013, *46*(2), 1410–1421.
38. Ostadi, A., Javanmard, A., Amani Machiani, M., Kakaei, K. Optimizing antioxidant activity and phytochemical properties of peppermint (Mentha piperita L.) by integrative application of biofertilizer and stress-modulating nanoparticles under drought stress conditions. *Plants*, 2022, *12*(1), 151.
39. Shenavaie Zare, A., Ganjeali, A., Vaezi Kakhki, M.R., Cheniany, M., Mashreghi, M. Plant elicitation and TiO_2 nanoparticles application as an effective strategy for improving the growth, biochemical properties, and essential oil of peppermint. *Physiology and Molecular Biology of Plants*, 2022, *28*(7), 1391–1406.
40. Aliakbarpour, F., Vaziri, A., Mohseni, M. Green synthesis of Au-Ag nanoparticles using Mentha piperita and effects of Au-Ag alloy nanoparticles on the growth of Mentha piperita under salinity stress. *Journal of Plant Process and Function*, 2020, *9*(37), 47–58.
41. Khanchi, S., Hashemi Khabir, S.H., Hashemi Khabir, S.H., Golmoghani Asl, R., Rahimzadeh, S. Foliar spray of magnesium oxide nanoparticles alleviates cadmium toxicity in mint (*Mentha crispa* L.). Available at SSRN 4500379.
42. Koleva, L., Umar, A., Yasin, N.A., Shah, A.A., Siddiqui, M.H., Alamri, S., Riaz, L., Raza, A., Javed, T., Shabbir, Z. Iron oxide and silicon nanoparticles modulate mineral nutrient homeostasis and metabolism in cadmium-stressed Phaseolus vulgaris. *Frontiers in Plant Science*, 2022, *13*, 806781.
43. Lu, M., Liang, Y., Lakshmanan, P., Guan, X., Liu, D., Chen, X. Magnesium application reduced heavy metal-associated health risks and improved nutritional quality of field-grown Chinese cabbage. *Environmental Pollution*, 2021, *289*, 117881.
44. Lafmejani, Z.N., Jafari, A.A., Moradi, P., Moghadam, A.L. Impact of foliar application of copper sulphate and copper nanoparticles on some morpho-physiological traits and essential oil composition of peppermint (L.). *Herba Polonica*, 2018, *64*(2), 13–24.

CHAPTER 6

Abiotic Stress Alleviation in Peppermint Using Nanoparticles and Depolymerized Polysaccharides

JAFAR FATHI QARACHAL,[1] EHSAN SHEIDAEE,[2] and POURYA BAZYAR[3,*]

[1]*Plant Pathology Department, Tarbiat Modares University, Tehran, Iran*

[2]*Biosystems Engineering Department, Tarbiat Modares University, Tehran, Iran*

[3]*Department of Mechanical Engineering and Production Management, Hamburg University of Applied Science, Hamburg, Germany*

*Corresponding author

ABSTRACT

Peppermint (*Mentha* × *piperita* L.) is a perennial plant belonging to the Lamiaceae family that produces creeping stolon with a strong aroma. In the current era, nanotechnology can be used in the cultivation and exploitation of various crops, including peppermint, to combat abiotic stresses that pose a big challenge to the growth and productivity of peppermint. Abiotic stresses resulting from water deficit, high salinity, or periods of drought adversely affect plant growth and development and represent major selective forces during plant evolution. Stress-reducing nanoparticles along with biofertilizers have been used to reduce the harmful effects of abiotic stresses and improve antioxidant activity and phytochemical properties of peppermint. Depolymerized polysaccharides have been studied for their potential to reduce abiotic stress in peppermint. Additionally, depolymerized compounds have been shown to enhance salt stress tolerance in plants.

Peppermint: Novel Strategies for Performance Improvement. Bilal Amhad, Tariq Ahmad Dar, and Khalid Rehman Hakeem (Eds.)

DOI: 10.1201/9781779641557-6

Radiologically depolymerized carrageenan has been found to reduce the harmful effects of drought stress and improve the biochemical characteristics of peppermint. Oligochitosan, a depolymerized form of chitosan, has been found to fortify antioxidative and photosynthetic metabolism and enhance secondary metabolite accumulation in arsenic-stressed peppermint. Overall, some nanoparticles and depolymerized polysaccharides have shown potential for controlling abiotic stress in peppermint and improving its growth.

6.1 INTRODUCTION

6.1.1 MENTHA PIPERITA

Peppermint (*M. piperita*) is a natural hybrid perennial herb of the Lamiaceae family, derived from water mint (*M. aquatic*) and spearmint (*M. spicata*). Since ancient Egyptian times, peppermint has been employed as a remedy for indigestion. Dried peppermint leaves, dating back to 1000 BC, were discovered in the pyramids of Egypt. The ancient Greeks and Romans utilized it as a stomach soother. In the 18th century, peppermint gained popularity as a folk remedy in Western Europe for addressing issues such as nausea, vomiting, morning sickness, respiratory infections, and menstrual disorders [1, 2]. The cultivation of peppermint in England began in the latter part of the 17th century. The leaves are usually used fresh as a culinary herb, and the dried flowers are used to flavor tea, desserts, drinks, salads, and other foods. Its essential oil is also used as a flavoring agent for adding taste or scent to foods, cosmetics, soaps, toothpaste, mouthwash, and other products, and peppermint extract has medicinal uses as well. Peppermint (*Mentha* × *piperita*) is a plant that produces creeping stems. It has square stems and smooth dark green leaves. The leaves are opposite, simple, and toothed, and can be up to 9 cm long, with a sweet, pungent smell and a warm and pungent taste with a cool aftertaste. The flowers are violet, measuring 6–8 mm in length, featuring a four-lipped corolla approximately 5 mm in diameter. They emerge in circular arrangements around the stem, creating dense, colorless clusters (Figure 6.1). The blooming period extends from mid to late summer. The rhizomes are dispersed and possess a fleshy nature, accompanied by fibrous roots. Chromosome count variability is observed, with recorded 2n counts of 66, 72, 84, and 120 [3, 4]. Plants face diverse abiotic stresses that can impede their growth and development, ultimately diminishing crop productivity. These stresses encompass fluctuations in temperature, inadequate or excessive water supply, elevated salinity, exposure to heavy metals, and ultraviolet radiation [5, 6]. These stressors can lead to changes in gene expression, cellular metabolism, shifts in growth rates, and alterations in crop yields. A plant's initial defense against abiotic stress is rooted in its roots. If the soil supporting the plant is both healthy and biologically diverse, the plant stands a better chance of surviving challenging conditions. Plants exhibit intricate and diverse responses to abiotic stresses, operating on multiple levels encompassing the whole plant, cellular processes, and molecular mechanisms. The specific response of plants to abiotic stress is contingent upon the affected tissue or organ. Various abiotic stresses affecting plants include drought, salinity, extreme temperatures, light variations, and pollution. Plants adapt to these stressors by modifying their structure and morphology, synthesizing osmolytes, neutralizing reactive oxygen species, and regulating stomatal opening. To cultivate crops that are resilient to abiotic stress and possess enhanced agronomic traits, it is essential to possess comprehensive knowledge of the mechanisms underlying plant adaptation to such stressors. Therefore, achieving a holistic understanding of the diverse responses linked to abiotic stress adaptation is crucial. This can be accomplished by leveraging various high-throughput tools such as proteomics, metabolomics, and transcriptomics [6–8].

FIGURE 6.1 *M. piperita* plants.

Peppermint, scientifically known as *M. piperita*, experiences a range of abiotic stresses that can curtail its growth, impede development, and decrease overall crop productivity. Factors like drought stress, heavy metals, salinity, and plant growth regulators influence peppermint. Research indicates that drought stress has adverse effects on the growth, productivity, and quality of peppermint. Interestingly, the same study highlights that mild drought stress can augment the essential oil (EO) content of peppermint, as well as enhance the concentrations of key EO constituents such as menthol, 1,8-cineole, and neo-menthol. An additional investigation revealed that peppermint exhibits promising molecular and biochemical tolerance to lead (Pb) stress. The reactions of peppermint to abiotic stresses are intricate and occur at multiple levels, contingent upon the specific tissue or organ affected by the stress. Diverse approaches are being explored to address abiotic stresses in peppermint, encompassing the application of biofertilizers, stress-modulating nanoparticles, and plant growth regulators [9–12].

Nanotechnology stands out as a cutting-edge and highly hopeful approach for addressing abiotic stress conditions in plants, including challenges like heavy metal stress, heat stress, drought stress, and salinity stress. Nanoparticles (NPs) represent a category of materials comprising particulate substances with at least one dimension measuring less than 100 nm. While the term typically refers to particles with diameters between 1 and 100 nm, it is occasionally extended to larger particles, reaching up to 500 nm, or to fibers and tubes with dimensions less than 100 nm in only two directions [13]. Nanoparticles prove effective in aiding plants to acclimate to challenging conditions, resulting in increased yields by mitigating the detrimental impacts of abiotic stress. The application of nanotechnology holds the potential to boost crop productivity by enhancing plant tolerance mechanisms in the face of abiotic stressors. Plants inherently possess sophisticated molecular programs to perceive stressors and adjust accordingly, and NPs can amplify these adaptive responses. Particularly in the context of temperature stress, NPs play a crucial role in enabling plants to endure abiotic stresses. The application of NPs in varying concentrations has been shown to enhance plant growth and hydration, mitigating the effects of heat stress. It is noteworthy that at lower concentrations, NPs exhibit antioxidative properties, whereas at higher concentrations, they may induce oxidative stress in plants [10, 14].

Polysaccharides are complex carbohydrates made up of long chains of monosaccharides. Depolymerized polysaccharides (DPPS) are smaller units of complex carbohydrates that are produced by breaking down long chains of polysaccharides. Depolymerization of polysaccharides refers to the process of breaking down these long chains into smaller units. This process can occur through various methods, including enzymatic depolymerization, oxidative depolymerization, and chemical depolymerization. Some of the applications of DPPS include fruit softening during ripening, bacterial polysaccharide breakdown, and lignin depolymerization. Enzymatic depolymerization is a common method used to break down polysaccharides, and it involves the use of enzymes to catalyze the process [15]. A comprehensive review of polysaccharide enzymatic depolymerization studies is presented in oxidative depolymerization: This entails utilizing reactive oxygen/nitrogen species for the breakdown of polysaccharides. The breakdown of these polysaccharides results in the relaxation of the cell wall, significantly contributing to the softening of fruits during the ripening process. Chemical depolymerization, a method involving the use of chemicals to break down polysaccharides, is pivotal in the valorization of lignocellulosic biomass. Cost-effective systems, including pure water or other solvents, can be employed for the chemical depolymerization process [16]. Numerous studies indicate that depolymerized chitosan, known as chitosan oligosaccharides (COS), has the potential to improve plant tolerance to abiotic stresses. Chitosan, a biopolymer derived from chitin, undergoes depolymerization to form COS. These oligosaccharides exhibit diverse biological activities, including antioxidant, anti-inflammatory, and immunomodulatory effects. The positive impact of COS extends to plant stress responses, where they enhance physiological reactions and alleviate the detrimental effects of abiotic stresses through stress transduction pathways mediated by secondary messengers. COS has demonstrated efficacy in enhancing salt tolerance in wheat seedlings and drought tolerance in maize plants. As a product of chitosan degradation, COS exhibits greater bioactivity compared to chitosan [17]. Alginate, a natural anionic polysaccharide extracted from brown seaweed, is a member of the linear copolymer family present in the cell wall matrix of brown seaweed. It consists of two monomeric units, namely β-(1,4) linked D-mannuronic acid (ManAp or M) and α-(1,4)-linked L-guluronic acid (GulAp or G). Depolymerized alginates, derived from brown seaweed, exhibit diverse applications in the food, pharmaceutical, and biomedical industries.

Notably, oligo–alginate coatings have been found to alleviate the adverse effects of salinity on plants. All fractions of oligo–alginates were found to promote plant growth, with the most pronounced effect observed in the fraction with a molecular weight of 32,000 g mol^{-1} [18]. Depolymerized carrageenan is a form of carrageenan that has been broken down into smaller units. Radiolytically depolymerized carrageenan has been found to have a proliferating effect on physiological attributes of some plants under drought stress conditions [19].

6.2 VARIOUS NPS EFFECT ON *M. PIPERITA* PHYSIOLOGY AND ABIOTIC STRESSES

Nanoparticles have emerged as key contributors in bolstering plant resilience to abiotic stress conditions. Numerous studies have demonstrated that employing NPs can aid plants in withstanding adverse conditions like drought, salinity, and temperature stress [20] (Figure 6.2).

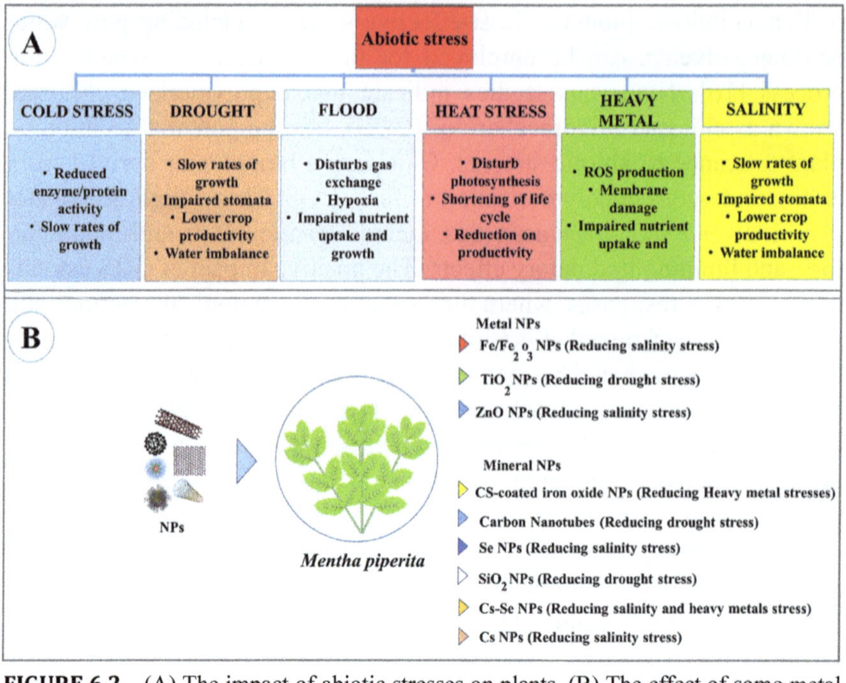

FIGURE 6.2 (A) The impact of abiotic stresses on plants. (B) The effect of some metal and mineral nanoparticles on the abiotic stress of peppermint.

6.2.1 METAL NPS

Metal nanoparticles are small entities on a submicron scale, composed of pure metals like gold, platinum, silver, titanium, zinc, cerium, iron, and thallium. These nanomaterials consist of a single element, existing either as individual atoms or clusters of multiple atoms. The fascination with metallic nanoparticles has spanned over a century among scientists, and their application has become widespread in biomedical sciences and engineering. Typically, these nanoparticles are formed through a "bottom-up" chemical process involving a metal salt and a potent reducing agent, such as sodium. Recent studies indicate that metal nanoparticles, including silver, gold, copper, nickel, platinum, titanium, and zinc nanoparticles, exhibit diverse biological applications [21, 22].

Metal nanoparticles have been found to have potential applications in alleviating abiotic stresses in plants. Studies have shown that nanoparticles can enhance plant tolerance mechanisms under abiotic stress conditions. Metal nanoparticles can contribute to mitigating abiotic stresses in plants through various mechanisms. These include enhancing tolerance to abiotic stress, preserving nutrient element homeostasis, promoting antioxidant enzymes, fortifying the antioxidative defense system, fostering plant growth, augmenting nutrient absorption, and influencing key metabolic cascades in plants [23, 24].

6.2.1.1 IRON (FE) NPS

Iron nanoparticles have demonstrated an ability to enhance the stress resistance of *M. piperita* when subjected to salinity stress. Recent studies indicate that iron oxide nanoparticles (Fe_2O_3 NPs) at concentrations ranging from 10 to 30 μM positively influence the growth of peppermint under salt stress. The application of Fe_2O_3 NPs resulted in increased fresh and dry weight of the leaves, along with elevated levels of phosphorus, potassium, iron, zinc, and calcium. Moreover, the use of Fe_2O_3 NPs led to a reduction in lipid peroxidation and proline content in the plants, while concurrently enhancing the activity of antioxidant enzymes such as catalase, superoxide dismutase, and guaiacol peroxidase. Consequently, it is proposed that carefully selected concentrations of iron nanoparticles can be employed to aid mint plants in resisting salinity stress [25].

6.2.1.2 TiO_2

This research has explored the potential of titanium dioxide nanoparticles (TiO_2 NPs) in promoting plant growth and enhancing the yield and quality of essential oils in medicinal and aromatic plants, including peppermint. The beneficial impacts of TiO_2 NPs on peppermint growth and essential oil production have been observed across various concentrations and application methods, such as foliar spraying and soil application. TiO_2 NPs can activate diverse defense mechanisms that contribute to plant tolerance against various abiotic stress factors. However, the effects of TiO_2 NPs may vary under different environmental conditions, in diverse plant species, and depending on the concentrations applied. It is crucial to emphasize that the size, shape, and concentration of TiO_2 NPs play significant roles in their application, and excessive concentrations may result in toxicity effects. [10].

6.2.1.3 ZNO NPS

Zinc oxide nanoparticles (ZnO NPs) have been demonstrated to positively impact plant tolerance to abiotic stress, particularly in drought and salinity stress. Applying ZnO NPs has been shown to enhance relative water content, membrane stability index, and antioxidant enzyme activity in plants experiencing drought stress. Additionally, ZnO NPs can mitigate the adverse effects of salinity stress on plant growth, reducing levels of malondialdehyde and hydrogen peroxide in plant tissues. Studies on *Brassica napus* indicate that Zn NPs promote growth and do not induce stress or harm to plants. It is crucial to recognize that the effects of ZnO NPs on plant growth and stress tolerance may vary based on concentration and application method. While ZnO NPs hold promise as a strategy for enhancing peppermint's tolerance to abiotic stress, further research is necessary to determine the optimal concentration and application method [26].

6.2.2 MINERAL AND NONMETALLIC NPS

Nonmetallic and mineral nanoparticles possess the capacity to alleviate abiotic stresses in plants by enhancing stress tolerance, preserving nutrient element balance, promoting antioxidant enzyme activity, and fostering plant growth. The observed effects of mineral nanoparticles extend to the modification of physiological, biochemical, and molecular mechanisms in plants under abiotic stresses, ultimately mitigating the impact of these stressors on plant well-being [20, 27].

6.2.2.1 SILICA DIOXIDE NANOPARTICLES (SIO_2 NPS)

SiO_2 nanoparticles have been explored for their potential to enhance plant growth and bolster tolerance to environmental stressors. These nanoparticles exhibit the ability to elevate chlorophyll content in peppermint leaves, thereby improving photosynthesis, enhancing overall plant growth, and increasing the antioxidant activity in peppermint. Moreover, SiO_2 nanoparticles play a role in mitigating the adverse effects of lead stress on plants. Additionally, they contribute to the augmentation of glandular trichomes in peppermint, responsible for producing essential oils and secondary metabolites, while also enhancing proline concentration and the activity of antioxidant enzymes. These effects collectively aid plants in overcoming environmental stress. However, the direct impact of SiO_2 nanoparticles on plant growth may vary-being positive, nonsignificant, or negative—depending on the concentration and plant species. Notably, SiO_2 nanoparticles have demonstrated efficacy in pest control, with reported lethality against pests such as *Callosobruchus maculatus*, offering potential benefits for crop protection. It is important to note that while SiO_2 nanoparticles present a sustainable approach for plant growth and crop protection, the use of nonselective insecticides may have adverse effects on both pests and beneficial organisms, including natural enemies that play a role in controlling pest populations [28, 29].

6.2.2.2 SE NPS

Limited research exists on the impact of selenium nanoparticles (Se NPs) on *M. piperita* (peppermint) under stress conditions. Nevertheless, some studies have explored the effects of Se NPs on other mint species facing salinity stress. One study indicated that Se NPs can mitigate salinity-induced damage and enhance the phytochemical characteristics of pineapple mint *(Mentha suaveolens Ehrh.)*. Another study demonstrated that Se

NPs could elevate the concentration of a specific compound in pineapple mint when subjected to moderate and mild salinity stresses. Furthermore, an investigation involving chitosan–selenium nanoparticles (Cs–Se NPs) revealed that treatments with Se and Cs–Se NPs could ameliorate the adverse impacts of Cd-stress conditions by enhancing agronomic traits. However, there is currently no research available on the effects of Se–NPs specifically on peppermint under drought stress conditions [30].

6.2.2.3 *CHITOSAN NPS*

Several studies have explored the impact of Cs NPs on different mint species facing drought stress. One study revealed that Cs NPs can alleviate drought stress in peppermint, enhancing both plant growth and various physiological parameters. Another investigation demonstrated that the application of chitosan-coated iron oxide nanoparticles (Fe–CTs NPS) increased the production of essential oils in peppermint when exposed to drought stress. Furthermore, nanoparticles have been recognized for their potential in addressing nutrient deficiencies, enhancing stress tolerance, and improving crop yield and quality. In a study investigating the antibiofilm properties of *M. piperita* essential oil-loaded chitosan nanoparticles (MPEO–CsNPs) against *Streptococcus mutans*, the researchers found that MPEO exhibited antibiofilm effects against the bacteria. However, as of now, there is no research available on the effects of Cs NPs specifically on peppermint under salinity stress conditions [31–33].

6.2.2.4 *CARBON NPS*

There is scarce research on the impact of carbon nanotubes (CNTs) on *M. piperita* (peppermint) in the context of abiotic stress conditions. Nonetheless, a singular study indicates that multiwalled carbon nanotubes can notably enhance both growth and salt tolerance in peppermint plants. This improvement is attributed to the promotion of water uptake, increased antioxidant activity, and enhanced phytochemical responses under conditions of salinity stress [34]. However, the effects of CNTs on *Mentha* sp. beyond salinity stress are not clear from the search results.

6.3 DPPS EFFECT ON M. PIPERITA AND OTHER MENTHA SPECIES PHYSIOLOGY AND ABIOTIC STRESSES

Derived from plants, DPPS have demonstrated favorable effects on plants facing abiotic stress conditions. These effects encompass the enhancement of salt stress tolerance, improvement in plant growth, and increased resistance to various abiotic stresses [35, 36].

6.3.1 CHITOSAN OLIGOSACCHARIDES

Chitosan oligosaccharides (COS) have been found to have a protective, biostimulating, and eliciting effect on *M. piperita* under stress conditions. The application of a chitosan solution in acetic acid as an elicitor in *M. piperita* suspension cultures resulted in a notable rise in menthol accumulation. Furthermore, subjecting *M. piperita* to chitosan treatment in greenhouse conditions led to an augmented total content of phenolic compounds and flavonoids, along with an enhancement in the antioxidant activity of the extracts. An additional study revealed that oligochitosan can strengthen both antioxidative and photosynthetic metabolism, leading to an increased accumulation of secondary metabolites in arsenic-stressed peppermint. Furthermore, COS have demonstrated their efficacy in enhancing artemisinin yield in plants exposed to well-watered and drought-stressed conditions. Concurrent application of depolymerized chitosan and GA3 has been shown to modulate photosynthesis, essential oil production, and menthol production in peppermint. Additionally, COS exhibit a biostimulating effect on the growth and essential oil profile of *M. arvensis*. Collectively, these findings suggest that COS can positively impact *M. piperita* under stress conditions by promoting the accumulation of secondary metabolites, enhancing antioxidant activity, and improving growth and photosynthetic metabolism [37–39].

6.3.2 DEPOLYMERIZED CARRAGEENAN

Carrageenans obtained from red seaweeds, after depolymerization, exhibit diverse impacts on both plant growth and tolerance to abiotic stress [40, 41]. Some of the key findings from the search results include:

Promotion of plant growth: Carrageenans, including depolymerized forms, have been found to promote plant growth in various species.

As an example, λ-carrageenan has been demonstrated to enhance the growth of banana plants through the improvement of cellular metabolism, nutrient absorption, and cellular homeostasis [42]. In a different investigation, the proliferative impact of carrageenan depolymerized through radiolysis was shown on physiological characteristics, plant water-related parameters, and essential oil production in *Cymbopogon flexuosus Steud* under conditions of drought stress [19]. Both gamma-irradiated carrageenan and depolymerized carrageenan exhibit beneficial effects on *Mentha arvensis* L. plants. Gamma-irradiated carrageenan has been observed to regulate physiological activities, active constituents, and essential oil production in this plant species. Meanwhile, depolymerized carrageenan enhances physiological activities and menthol production in *Mentha arvensis* L. [43, 44]. The results imply that both gamma-irradiated and depolymerized carrageenans hold promise as plant biostimulants for enhancing crop production and quality across diverse environmental conditions.

Boosting defense mechanisms: Oligosaccharides derived from carrageenan have been discovered to promote plant defenses against both biotic and abiotic stresses. These impacts are mediated through distinct signaling pathways, with specific oligosaccharides exhibiting varied effects. For instance, oligoalginates induce protection against pathogens through a salicylic acid-dependent pathway, address abiotic stress via an abscisic acid-dependent pathway, and influence growth through a mechanism that is currently unidentified [35, 45].

Improvement of plant water relations (Drought stress): The depolymerized form of carrageenan has demonstrated a stimulative impact on plant water relation parameters, presenting potential benefits for plants experiencing drought stress conditions [19].

6.3.3 DEPOLYMERIZED PECTIN OLIGOSACCHARIDES

Pectin constitutes a group of glycans and glycoconjugates within the plant cell wall, playing crucial roles in plant development and responses to both biotic and abiotic stress. Pectin oligosaccharides, specifically oligogalacturonides, participate in plant reactions to abiotic stresses like saline stress, osmotic stress, and temperature variations. The significance of pectin-*O*-acetylation in plant growth and development is acknowledged, although there is limited understanding of its regulatory mechanisms in key processes within the plant cell wall [46–48]. It has been documented that pectin oligogalacturonides trigger resistance in *Arabidopsis thaliana* against *Pseudomonas syringae pv. Tomato via* salicylic acid-mediated signaling pathways. [49]. Moreover, pectin oligogalacturonides have been demonstrated to bolster plant tolerance to abiotic stress, including salt stress, by elevating the activity of antioxidant enzymes and modulating the expression of stress-responsive genes [46]. Under cold stress conditions, wall-associated kinases can attach to either oligogalacturonides or pectin within the cell wall. This binding can result in the activation of pectin-induced defense responses, suggesting that the cell wall receptors may have a potential role as a cold stress sensor by means of oligogalacturonide reception. Variations in the activities of enzymes that modify pectin have been observed in various species under diverse growth conditions in reaction to cold stress [50]. As elicitors, oligogalacturonides might main factor in triggering cold resistance in plants of the *Mentha* spp.

6.3.4 DEPOLYMERIZED ALGINATE

Information is scarce regarding the precise impacts of depolymerized alginate on *M. piperita* under abiotic stress. Nevertheless, certain studies have explored the effects of sodium alginate and irradiated sodium alginate on other plants, such as *M. arvensis* [51], *Foeniculum vulgare* Mill. [52], and *Vigna radiata* L. [53], under different conditions such as heavy metal stress (cadmium stress) and drought stress. These investigations indicate that sodium alginate, especially in the form of depolymerized oligosaccharides, can exhibit a stimulating impact on both plant growth and physiological activities. Therefore, it is possible that depolymerized alginate could have similar effects on *M. piperita* under abiotic stress, but further research is needed to confirm this.

6.4 CONCLUSION

Nanoparticles, both metallic and nonmetallic, and depolymerized polysaccharides have been found to be effective in reducing abiotic stresses of plants. The use of nanoparticles can lead to better fertility, more essential oil production, and improvement of morphological and physiological characteristics of plants, including the genus *Mentha* sp. and its famous species, such as peppermint, which have been used as medicinal plants for centuries. Nanoparticles play a crucial role in nutrient deficiencies, enhancing stress tolerance, and elevating crop yield and quality. The utilization of nanomaterials in agriculture holds paramount significance for minimizing nutrient losses, managing pests, and augmenting stress resistance. Nevertheless, further research is essential to explore the impact of nanoparticles and depolymerized polysaccharides in mitigating the effects of abiotic stresses on peppermint.

KEYWORDS

- **peppermint**
- **Lamiaceae**
- **abiotic stresses**
- **nanoparticle**
- **depolymerized polysaccharide**

REFERENCES

1. Maspero, G. *Études de mythologie et d'archéologie égyptiennes*. E. Leroux. 1893.
2. Elansary, H.O., Ashmawy, N.A. Essential oils of mint between benefits and hazards. *Journal of Essential Oil Bearing Plants*, 2013, *16*(4), 429–438.
3. Harley, R.M., Brighton, C.A. Chromosome numbers in the genus *Mentha* L. *Botanical Journal of the Linnean Society*, 1977, *74*(1), 71–96.
4. Malekmohammad, K., Rafieian-Kopaei, M., Sardari, S., Sewell, R. D. Toxicological effects of *Mentha* x *piperita* (peppermint): A review. *Toxin Reviews*, 2021, *40*(4), 445–459.
5. Dey, S., Raichaudhuri, A. Abiotic stress in plants. In: *Advances in Plant Defense Mechanisms*. IntechOpen. 2022.
6. Zhang, H., Zhu, J., Gong, Z., Zhu, J.K. Abiotic stress responses in plants. *Nature Reviews Genetics*, 2022, *23*(2), 104–119.
7. He, M., He, C.Q., Ding, N.Z. Abiotic stresses: General defenses of land plants and chances for engineering multistress tolerance. *Frontiers in Plant Science*, 2018, *9*, 1771.

8. Petrov, V., Gechev, T. ROS and abiotic stress in plants 2.0. *International Journal of Molecular Sciences*, 2023, *24*(15), 11917.
9. Fathi, A., Oveysi, M., Nasri, M., Tohidi, H., Kasraei, P. Physiological responses of peppermint (*Mentha piperita* L.) to plant growth regulators and salinity stress. *Iranian Journal of Plant Physiology*, 2020, *11*(1), 3499–3508.
10. Ostadi, A., Javanmard, A., Amani Machiani, M., Kakaei, K. Optimizing antioxidant activity and phytochemical properties of peppermint (*Mentha piperita* L.) by integrative application of biofertilizer and stress-modulating nanoparticles under drought stress conditions. *Plants*, 2022, *12*(1), 151.
11. Chiappero, J., Cappellari, L.D.R., Palermo, T.B., Giordano, W., Banchio, E. Influence of drought stress and PGPR inoculation on essential oil yield and volatile organic compound emissions in *Mentha piperita*. *Horticulturae*, 2022, *8*(12), 1120.
12. Safarian, M., Nodehi, M.S., Bihamta, M.R., Sobhanverdi, S. *Mentha piperita* L demonstrates a potential tolerance at molecular and biochemical level to Pb stress: New perspectives. *South African Journal of Botany*, 2023, *159*, 635–641.
13. Rodrigues, M.P., Pinto, P.N., Dias, R.R.D.S., Biscoto, G.L., Salvato, L.A., Millán, R.D.S., Keller, K.M. et al. The antimicrobial applications of nanoparticles in veterinary medicine: A comprehensive review. *Antibiotics*, 2023, *12*(6), 958.
14. Al-Khayri, J.M., Rashmi, R., Surya Ulhas, R., Sudheer, W.N., Banadka, A. and Nagella, P., et al. The role of nanoparticles in response of plants to abiotic stress at physiological, biochemical, and molecular levels. *Plants*, 2023, *12*(2), 292.
15. Rollings, J. Enzymatic depolymerization of polysaccharides. *Carbohydrate Polymers*, 1985, *5*(1), 37–82.
16. Song, T., Cai, W., Wang, F., Lv, G. Effects of different depolymerisation methods on the physicochemical and antioxidant properties of polysaccharides derived from Sparassis latifolia. *Process Biochemistry*, 2021, *110*, 110–117.
17. Li, J., Han, A., Zhang, L., Meng, Y., Xu, L., Ma, F., Liu, R. Chitosan oligosaccharide alleviates the growth inhibition caused by physcion and synergistically enhances resilience in maize seedlings. *Scientific Reports*, 2022, *12*(1), 162.
18. Salachna, P., Grzeszczuk, M., Meller, E., Soból, M. Oligo-alginate with low molecular mass improves growth and physiological activity of Eucomis autumnalis under salinity stress. *Molecules*, 2018, *23*(4), 812.
19. Singh, M., Khan, M.M.A., Uddin, M., Naeem, M., Qureshi, M.I. Proliferating effect of radiolytically depolymerized carrageenan on physiological attributes, plant water relation parameters, essential oil production and active constituents of Cymbopogon flexuosus Steud. under drought stress. *PLoS One*, 2017, *12*(7), e0180129.
20. El-Saadony, M.T., Saad, A.M., Soliman, S.M., Salem, H.M., Desoky, E.S.M., Babalghith, A.O., Elrys, A.S., El-Tarabily, K.A., AbuQamar, S.F. Role of nanoparticles in enhancing crop tolerance to abiotic stress: A comprehensive review. *Frontiers in Plant Science*, 2022, *13*, 946717.
21. Mody, V.V., Siwale, R., Singh, A., Mody, H.R. Introduction to metallic nanoparticles. *Journal of Pharmacy and Bioallied Sciences*, 2010, *2*(4), 282.
22. Paulami, D.A.M., Paret, M.L., Mondal, R., Mandal, A.K. Advancement of noble metallic nanoparticles in agriculture: A promising future. *Pedosphere*, 2023, *33*(1), 116–128.

23. Yaqoob, A.A., Ahmad, H., Parveen, T., Ahmad, A., Oves, M., Ismail, I.M., Qari, H.A., Umar, K., Mohamad Ibrahim, M.N. Recent advances in metal decorated nanomaterials and their various biological applications: A review. *Frontiers in Chemistry*, 2020, 8, 341.
24. Landa, P. Positive effects of metallic nanoparticles on plants: Overview of involved mechanisms. *Plant Physiology and Biochemistry*, 2021, 161, 12–24.
25. Askary, M., Talebi, S.M., Amini, F., Bangan, A.D.B. Effects of iron nanoparticles on *Mentha piperita* L. under salinity stress. *Biologija*, 2017, *63*(1).
26. Mogazy, A.M., Hanafy, R.S. Foliar spray of biosynthesized zinc oxide nanoparticles alleviate salinity stress effect on Vicia faba plants. *Journal of Soil Science and Plant Nutrition*, 2022, *22*(2), 2647–2662.
27. Kumari, V.V., Banerjee, P., Verma, V.C., Sukumaran, S., Chandran, M.A.S., Gopinath, K.A., Venkatesh, G., Yadav, S.K., Singh, V.K., Awasthi, N.K. Plant nutrition: An effective way to alleviate abiotic stress in agricultural crops. *International Journal of Molecular Sciences*, 2022, *23*(15), 8519.
28. Wei, C., Zhang, Y., Guo, J., Han, B., Yang, X., Yuan, J. Effects of silica nanoparticles on growth and photosynthetic pigment contents of Scenedesmus obliquus. *Journal of Environmental Sciences*, 2010, *22*(1), 155–160.
29. Ahmad, B., Khan, M.M.A., Jaleel, H., Shabbir, A., Sadiq, Y., Uddin, M. Silicon nanoparticles mediated increase in glandular trichomes and regulation of photosynthetic and quality attributes in *Mentha piperita* L. *Journal of Plant Growth Regulation*, 2020, *39*, 346–357.
30. Kiumarzi, F., Morshedloo, M.R., Zahedi, S.M., Mumivand, H., Behtash, F., Hano, C., Chen, J.-T., Lorenzo, J.M. Selenium nanoparticles (Se-NPs) alleviates salinity damages and improves phytochemical characteristics of pineapple mint (Mentha suaveolens Ehrh.). *Plants*, 2022, *11*(10), 1384.
31. Hidangmayum, A., Dwivedi, P., Katiyar, D., Hemantaranjan, A. Application of chitosan on plant responses with special reference to abiotic stress. *Physiology and Molecular Biology of Plants*, 2019, *25*, 313–326.
32. Giglou, M.T., Giglou, R.H., Esmaeilpour, B., Azarmi, R., Padash, A., Falakian, M., Sliwka, J., Gohari, G., Lajayer, H.M. A new method in mitigation of drought stress by chitosan-coated iron oxide nanoparticles and growth stimulant in peppermint. *Industrial Crops and Products*, 2022, 187, 115286.
33. Attaran Dowom, S., Karimian, Z., Mostafaei Dehnavi, M., Samiei, L. Chitosan nanoparticles improve physiological and biochemical responses of Salvia abrotanoides (Kar.) under drought stress. *BMC Plant Biology*, 2022, *22*(1), 364.
34. Fallah, R., Gerami, M., Ramezani, M. Beneficial role of multi-walled carbon nanotubes on physiological and phytochemical responses of *Mentha piperita* L under salinity stress. *Journal of Essential Oil Bearing Plants*, 2023, *26*(2), 323–342.
35. Ahmad, B., Jahan, A., Sadiq, Y., Shabbir, A., Jaleel, H., Khan, M.M.A. Radiation-mediated molecular weight reduction and structural modification in carrageenan potentiates improved photosynthesis and secondary metabolism in peppermint (Mentha piperita L.). *International Journal of Biological Macromolecules*, 2019, 124, 1069–1079.
36. Salachna, P. Effects of depolymerized gellan with different molecular weights on the growth of four bedding plant species. *Agronomy*, 2020, *10*(2), 169.

37. da Silva, E.A., Silva, V.N.B., de Alvarenga, A.A., Bertolucci, S.K.V. Biostimulating effect of chitosan and acetic acid on the growth and profile of the essential oil of Mentha arvensis L. *Industrial Crops and Products*, 2021, *171*, 113987.
38. Ahmad, B., Jaleel, H., Shabbir, A., Khan, M.M.A., Sadiq, Y. Concomitant application of depolymerized chitosan and GA3 modulates photosynthesis, essential oil and menthol production in peppermint (*Mentha piperita* L.). *Scientia Horticulturae*, 2019, *246*, 371–379.
39. Malerba, M., Cerana, R. Chitin-and chitosan-based derivatives in plant protection against biotic and abiotic stresses and in recovery of contaminated soil and water. *Polysaccharides*, 2020, *1*(1), 21–30.
40. Sara, K., Hossein, A., Masoud, S.J., Hassan, M. Effects of water deficit and chitosan spraying on osmotic adjustment and soluble protein of cultivars castor bean (Ricinus communis L.). *Journal of Stress Physiology & Biochemistry*, 2012, *8*(3), 160–169.
41. Naeem, M., Idrees, M., Aftab, T., Alam, M.M., Khan, M., Uddin, M., Varshney, L. Radiation Processed Carrageenan Improves Plant Growth, Physiological Activities, and Alkaloids Production in *Catharanthus roseus* L. *Advances in Botany*, 2015, *2015*.
42. Thye, K.L., Wan Abdullah, W.M.A.N., Balia Yusof, Z.N., Wee, C.Y., Ong-Abdullah, J., Loh, J.Y., Cheng, W.-H., Lamasudin, D.U., Lai, K.S. λ-Carrageenan promotes plant growth in banana via enhancement of cellular metabolism, nutrient uptake, and cellular homeostasis. *Scientific Reports*, 2022, *12*(1), 19639.
43. Naeem, M., Aftab, T., Idrees, M., Singh, M., Ali, A., Khan, M.M.A., Uddin, M., Varshney, L. Modulation of physiological activities, active constituents and essential oil production of *Mentha arvensis* L. by concomitant application of depolymerised carrageenan, triacontanol and 28-homobrassinolide. *Journal of Essential Oil Research*, 2017, *29*(2), 179–188.
44. Naeem, M., Idrees, M., Aftab, T., Khan, M.M.A., Varshney, L. Depolymerised carrageenan enhances physiological activities and menthol production in Mentha arvensis L. *Carbohydrate Polymers*, 2012, *87*(2), 1211–1218.
45. Sangha, J.S., Khan, W., Ji, X., Zhang, J., Mills, A.A., Critchley, A.T., Prithiviraj, B. Carrageenans, sulphated polysaccharides of red seaweeds, differentially affect *Arabidopsis thaliana* resistance to *Trichoplusia ni* (Cabbage Looper). *Plos One*, 2011, *6*(10), e26834.
46. Shin, Y., Chane, A., Jung, M., Lee, Y. Recent advances in understanding the roles of pectin as an active participant in plant signaling networks. *Plants*, 2021, *10*(8), 1712.
47. Le Gall, H., Philippe, F., Domon, J.M., Gillet, F., Pelloux, J., Rayon, C. Cell wall metabolism in response to abiotic stress. *Plants*, 2015, *4*(1), 112.
48. Shahin, L., Zhang, L., Mohnen, D., Urbanowicz, B.R. Insights into Pectin O-acetylation in plant cell wall: Structure, synthesis, and modification. *The Cell Surface*, 2023, *9*, 100099.
49. Ridley, B.L., O'Neill, M.A., Mohnen, D. Pectins: Structure, biosynthesis, and oligogalacturonide-related signaling. *Phytochemistry*, 2001, *57*(6), 929–967.
50. Wu, H.C., Bulgakov, V.P., Jinn, T.L. Pectin methylesterases: Cell wall remodeling proteins are required for plant response to heat stress. *Frontiers in Plant Science*, 2018, *9*, 1612.
51. Naeem, M., Idrees, M., Aftab, T., Khan, M.M.A., Varshney, L. Irradiated sodium alginate improves plant growth, physiological activities and active constituents in

Mentha arvensis L. *Journal of Applied Pharmaceutical Science*, 2012, *2*(Issue), 28–35.

52. Adeeba, S., Naeem, M., Shafia, N., Mohd, I., Tariq, A., Nadeem, H., Khan, M.M.A., Moinuddin, Lalit, V. An evaluation of the effects of irradiated sodium alginate on the growth, physiological activities and essential oil production of fennel (*Foeniculum vulgare* Mill.). *Journal of Medicinal Plants Research*, 2011, *5*(1), 15–21.
53. Moussa, H.R., Taha, M.A., Dessoky, E.S., Selem, E. Exploring the perspectives of irradiated sodium alginate on molecular and physiological parameters of heavy metal stressed *Vigna radiata* L. plants. *Physiology and Molecular Biology of Plants*, 2023, *29*(3), 447–458.

CHAPTER 7

Influence of Radiation-Processed Polysaccharides on Performance of Peppermint

AJMAT JAHAN,[1] M. MASROOR A. KHAN,[2] MOHD GULFISHAN,[3*] BILAL AHMAD,[4] KHAN BILAL MUKHTAR AHMED,[5] and ASFIA SHABBIR[2]

[1]*Department of Botany, Aligarh Muslim University, Aligarh, Uttar Pradesh, India*

[2]*Advanced Plant Physiology Section, Department of Botany, Aligarh Muslim University, Aligarh, Uttar Pradesh, India*

[3]*School of Agriculture & Technology Maya Devi University, Selaqui, Dehradun, Uttarakhand*

[4]*Department of Botany, Government Degree College Khansahib, Budgam, University of Kashmir, Jammu & Kashmir, India*

[5]*Faculty of Agricultural Sciences, Aligarh Muslim University, Aligarh, Uttar Pradesh, India*

*Corresponding author

ABSTRACT

Recognized as one of the earliest medicinal and aromatic plants of global economic importance, peppermint (*Mentha piperita* L.) is a hybrid of *Mentha aquatica* and *Mentha spicata*, chiefly valued for its essential oil containing a high amount of monoterpenes—menthol, menthone, and menthyl acetate. Essential oils in peppermint induce a wide range of uses in the pharmaceutical, cosmetics, food, and aromatherapy industries.

Peppermint: Novel Strategies for Performance Improvement. Bilal Amhad, Tariq Ahmad Dar, and Khalid Rehman Hakeem (Eds.)

DOI: 10.1201/9781779641557-7

Peppermint, cultivated in different areas of India, with Uttar Pradesh as the major center of cultivation. Enhancing peppermint yield and oil quality has been a major priority due to increased industrial demand. In recent years, radiation-processed polysaccharides (RPPs) derived from marine sources such as carrageenan, chitosan, and sodium alginate have emerged as effective stimulants for improving plant growth, physiology, and secondary metabolism. Radiation processing reduces polysaccharide molecular weight into bioactive oligomers that stimulate photosynthesis, nutrient uptake, enzymatic activity, antioxidant defenses, or essential oil accumulation when applied foliarly. Experimental evidence found that irradiated chitosan, sodium alginate, and carrageenan experiencing foliar application resulted in significantly enhanced peppermint growth characteristics, chlorophyll content, enzyme activities, and essential oil yield, with significantly increased menthol content. RPPs therefore have a lot of potential to increase peppermint yield and medicinal value while maintaining environmental sustainability

7.1 INTRODUCTION

One of the first documented fragrant and therapeutic plants in the Lamiaceae family is peppermint (*Mentha piperita* L.). *M. piperita* L. is derived from the Greek term Mintha [1], which refers to a mythical nymph said to have transformed into a plant, and the Latin piper, which refers to pepper. According to some researchers, piperita is derived from the Sanskrit word "*papali*," which means pepper. The father of taxonomy, Carolus Linnaeus, regarded peppermint as a species when he first described it in 1753 using specimens he had brought back from England. According to an investigation conducted by a team of USDA, NRCS experts [2], the plant *M. piperita* L. is a genetic hybrid derived when *Mentha aquatica* (water mint) crossed with *Mentha spicata* (spearmint).

The Mediterranean-born plant, which is native to Europe, can occasionally be discovered alongside its ancestor species [3]. It has spread to USA, New Zealand, Australia, the Galapagos Islands, and the United States. According to Akdogan et al. [4], the plant is widely cultivated in every part of the world. A large portion of India's peppermint production—roughly 90% of the country's total production—comes from the Uttar Pradesh areas (Moradabad, Rampur, Sambhal, and Barabanki districts), with the other 10% coming from the states of Haryana, Himachal Pradesh, and Punjab [5].

Peppermint is one of the first recognized plants for medicinal purposes, and was named the year's top therapeutic plant in 2004 [6]. Another name for this plant is "Hierbabuena," which translates to "good herb" (Figure 7.1). Due to the plant's use in food, confections, mouthwashes, chewing gum, toothpastes, herbal remedies, and the making of different scent therapy goods, it has significant economic significance. The oil of peppermint can be administered topically against antibacterial, rubefacient, antioxidant, antipruritic, antiseptic, and monastic effects [7–11]. It is also employed as a food stabilizer and as a treatment for headaches, migraines, nerve pain, and muscle pain [12–15]. According to Mohsenzadeh [16], peppermint can also be used to treat viral infections. The plant's radioprotective properties and the efficacy of its oil for cancer patients serve as further indications of its therapeutic significance [17].

FIGURE 7.1 *Mentha piperita* L. (peppermint) plants.

Because of its aromatic, antibacterial, and antifungal properties, the main ingredient in the oil, menthol, is used in many different products, together with those for oral hygiene, food, beauty products, and fragrance and essence industries, among others. As a result, peppermint oil ranks highly in regard to revenue from sales [18, 19]. Menthone and menthyl acetate, two more significant monoterpenes found in the oil of peppermint in addition to menthol, are employed for a variety of flavoring and cosmetic applications.

7.2 BOTANICAL DEPICTION

With squared stems, mint green opposing decussate, elongated, and elliptical leaves with a sharp tip and roughly dentate edges, peppermint is a perennial, evergreen, winter-hardy, rhizomatous, fast-flourishing plant that typically grows 30–90 cm tall. Three different trichome types can be seen on the leaves: (1) multicellular, nonglandular, simple hairs [3, 20], peltate glandular trichomes (PGTs), with an apical disc of eight secretory cells, and (2) trichomes that are small, capitate glandular and have a single ductless head cell. According to Glenn et al. [21], the PGTs, which serve as the primary locations for essential oil (EO)'s biosynthesis, are where the maximum of the peppermint EO's monoterpenes are created and stored [22] (Figures 7.2a and b). The root systems of the succulent rhizomes are fibrous. According to Neeraj et al. [23], the blooms are around 6–8 mm long and have a four-lobed corolla that emerges in verticillasters around the stem to form thick, blunt purple spikes.

FIGURE 7.2 (a) *Mentha piperita* plants. (b) Scanning electron microscope (SEM) image of *peppermint* leaf showing essential oil accumulating sites or secretory structures called as peltate glandular trichomes (PGTs).

7.2.1 MAIN CONSTITUENTS OF THE OIL

Peppermint oil is a combination consisting roughly 52% of monoterpenes and 9% of sesquiterpenes. However, additional groups, together with (9%) aromatic hydrocarbons, (9%) aldehydes, (7%) lactones, (8%) miscellaneous, and (8%) alcohols, have been demonstrated to exist in smaller amounts. Menthol dominates monoterpenes, accounting for 35–60% of the total. Menthone, menthyl acetate, 1,8-cineole (eucalyptol), menthofuran, isomenthone, neomenthol, and limonene follow closely behind at 2–44%, 0.7–23%, and 0.1–6%, respectively. A significant sesquiterpene is *β*-caryophyllene, with 1.6–1.8%. According to Alankar [24], Riachi and Maria [25], and other sources, the ratio of cineole to limonene in oil often stays at or below two.

7.2.2 USES OF MENTHA PIPERITA L.

Interest in *M. piperita* L. has always been high. due to its enormous pharmacological importance as well as its application in food and beauty products. Its medicinal applications include anti-inflammatory, depurative, pain-relieving, stimulating, emmenagogue, diaphoretic, antiemetic, antispasmodic, and anticatarrhal, in addition to its extensive use in food, confections, mouthwashes, chewing gums, toothpastes, herbal formulations, and multiple products for aromatherapy [9, 11, 26]. Additionally, it can be used to treat anorexia, pneumonia, diarrhea, colitis with ulcers, nausea, and liver issues. In addition to being utilized to treat migraines, headaches, nerve pain, and myalgia, peppermint oil can also be administered topically for antiseptic, antipruritic, rubefacient, antibacterial, and severe reasons [12, 27, 28]. Viral infections have also been treated using *M. piperita* L. The plant's water-based extract, as well as menthol, the primary component of its oil, have demonstrated promising efficacy toward herpes, influenza, and other viruses [16]. The plant's and its oil's radio-protective properties for cancer patients serve as further evidence of the medicinal significance [17]. Since monoterpenes make up the majority of peppermint oil, this plant can be used as a model for research on monoterpene metabolism [29]. The majority of the vital oils and resins in some plants are composed of multiple categories of volatile C-10 isoprenoids (monoterpenes), which also have a crucial chemo-ecological function in how the plants deal with abiotic and biotic components. While others, like menthol, which is the predominant monoterpene of the peppermint and corn mint EO, is of significant pharmaceutical relevance since it is utilized toward influenza, herpes, and other associated viruses, a few of these monoterpenes have been recognized as allelopathic agents, having a key part in defense toward pathogens, herbivorous and insects, etc. (Figure 7.3) [30, 31].

FIGURE 7.3 Diagrammatic representations of various uses of peppermint.

7.3 MARINE POLYSACCHARIDES

About half of all biological species on Earth are aquatic, including bacteria, microalgae, and seaweed. As a result, the oceans are a rich source of useful substances made by these biotic systems, such as different polysaccharides (RPPs) [32–35], polyunsaturated fatty acids [36], vitamins, antioxidants, enzymes, and bioactive peptides [37–39].

Therefore, marine organisms—specifically, agar, alginate, fucoidan, carrageenan, chitosan, and chitin—represent a rich and varied source of different RPPs. Due to their many qualities, like their biocompatibility, biodegradability, bioactivity, and nontoxicity, these RPPs have drawn a lot of interest and have been thoroughly studied in the fields of biotechnology, medicine, agriculture, pharmaceuticals, and biomedical fields [40].

These are unique, highly reactive compounds that create thermo-reversible gels; as such, they are perfect for encoding biological information. Its use is most likely to increase with structural modifications in RPPs, such as grafting, cross-linking, or degradation. Technology related to radiation is a helpful instrument to bring about these changes. In contrast to chemical and enzymatic hydrolysis, radiation technology breaks down RPPs in a single, clean phase. When these marine RPPs are exposed to ionizing radiation, their long-chain backbone breaks down into smaller oligomers, which significantly lowers their molecular weight. These radiations also help RPPs undergo operational alterations, like the addition of double bonds or more polar end groups, which are thought to be the cause of the oligomers' induction of important physiological and biological activities. When applied to plants using a variety of techniques, such as foliar sprays, these oligomers can stimulate a wide range of bioactivities pertaining to heavy metal stress reduction, phytoalexin induction, germination of seed, growth of shoot and root, production of flower, antimicrobial and antiviral activity, and the quantity and yield of vital oils in plants [41–48]. Furthermore, research claiming that the foliar application of radiation-processed RPPs fallouts in the amendment of active components in several aromatic and medicinal plants validates the well-known agricultural use of RPPs [49–61].

7.3.1 CARRAGEENAN

Carrageenans are a spontaneously occurring, economically valuable class of hydrophilic RPPs produced by the extraction of multiple kinds of marine red algae, the Rhodophyceae [62–64]. These are mixes of linear, highly sulfated, hydrophilic, high molecular weight molecules having galactose backbones. The word "carrageenan" is derived from the Irish term used to describe this seaweed, carrageen, which means "little rock." Since 1810, Ireland has been aware of the water-based extraction of red seaweeds to produce such a hydrophilic colloidal solution [65, 66]. Rhodophyta (the red algae) includes all types of seaweed that harvest carrageenan as their main component of cell wall, specifically *Euchema, Gigartina stellate, Solieria, Agardhiella, Chondrus crispus, Iridaea, Hypnea,* and *Sarconema,* etc [67–70]. They are mostly composed of 3- and 4-linked-D-galactopyranose (G-units and D-units, respectively) or 4-linked 3,6-anhydro-D-galacto-pyranose (DA-units) in alternation, creating carrageenan's disaccharide repetitive unit [71–74].

7.3.2 CHITOSAN

Chitosan, popularly recognized as poly-β-1/4-2-amino-2-deoxy-D-glucopyranose [75], is one of the utmost prevalent naturally occurring polymers on the planet following cellulose [76]. It is a copolymer made up of *N*-acetylglucosamine and glucosamine connected by α-(1→4) linkage. This is a biodegradable, nontoxic, and biocompatible straight RPP that is often derived from the cell walls of crustaceans, including crabs, prawns, lobsters, insects, and other creatures with exoskeletons [77]. Although it is also found in bacterial and fungal cell walls, it is typically not isolated from those structures [78]. Commercially manufactured chitosan has a mean molecular weight that ranges from 3800 to 20,000 Da. A few degrading techniques have been developed that decrease its molecular weight while not altering its chemical structure [79]. A relatively easy and environmentally benign method that has been successful in the mass manufacturing of oligo-chitosan is degradation brought on by radiation [80]. Chitosan has reaped a lot of attention owing to its intriguing qualities in the sectors of agriculture, biomedicine, and the environment [82, 83]. Chitosan and its by-products are widely employed in biotechnological applications, wastewater treatment, pharmaceuticals, and other industries for a variety of functions [84]. Chitosan is employed in agriculture for a variety of purposes, including encasing seeds, fruits, and vegetables [85], improving the development and production of crops, protecting crops from infectious agents [86, 87], and ensuring regulated and acceptable discharge chitosan promotes the generation of phytoalexins [88], reactive oxygen species [89], jasmonic acid biosynthesis [90], the conveying of novel initial receptive and defense-associated genes [91], and the synthesis of reactive oxygen species.

7.3.3 SODIUM ALGINATE

A naturally existing group of straight copolymers containing mannuronic acid and guluronic acid deposits is known as alginates, and also known as linear polyuronic acid hydrocolloids [92]. It is primarily produced by algae, particularly the brown algae (Phaeophyceae), a family that includes seaweeds like *Macrocystis pyrifera*, *Sargassum sinicola*, *Ascophyllum nodosum*, and different kinds of Laminaria. Alginate is also present in the capsules of some soil bacteria, including *Azotobacter*, *Pseudomonas*, etc. Alginates are straight homopolymeric blocks from a chemical perspective, but those that have been recovered from seaweeds are not random copolymers; rather, they are composed of blocks of identical and rigidly alternating residues with various structural inclinations and behavior. The empirical formula for sodium alginate ($NaC_6H_7O_6$) is composed of monomeric units of D-mannuronic acid and L-guluronic acid, which are linked by α-1→4 glycosidic linkages.

7.4 IMPACT OF RADIATION-PROCESSED POLYSACCHARIDES (CARRAGEENAN, CHITOSAN, AND SODIUM ALGINATE) ON PEPPERMINT

An innovative recent method for boosting plant yields involves radiolytically breaking down normal RPPs into trivial oligomers and then applying these aqueous solutions to plants. Plant growth regulators are a suitable option for raising the yield of economically significant end-products in plants because of their well-known growth-stimulating and terpenoid biosynthesis actions.

Pourhadi [93] reported that, under greenhouse conditions, chitosan, along with other biostimulants, increases the leaf and stem dry weight, EO content, and menthol content. The aforementioned values, however, were discovered to be much greater in the field than in the greenhouse.

Ahmad et al. [57] examined the effect of foliar application of diverse concentrations of radiation-treated chitosan on peppermint plants. Results displayed that the foliar-applied irradiated chitosan (ICN) meaningfully amended the growth, physiology, and yield of peppermint. Plant's dry and fresh weights grew by 65.3% and 62.7%, respectively, due to foliar administration of ICN with a concentration of 80 mg L^{-1}. In addition, the activities of carbonic anhydrase and nitrate reductase both rose by 22.6% and 19.7%, respectively, whereas chlorophyll fluorescence improved by 16.3% compared to the untreated plants. The amount of nitrogen in the leaves was 18.2% more than in the control group, while the totals of potassium and phosphorus rose by 7.9% and 10.1%, respectively. Forty-five days after planting, there was an 18.1% increase in phenol content. Treatment of 80 mg L^{-1} ICN increased the EO content and yield by 36.8% and 126.1%, respectively. According to gas chromatography, the amount of menthol rose by 14.8%, whereas menthone and menthyl acetate contents were decreased by 2.5% and 4.8% with 80 mg L^{-1} ICN.

A pot experiment was planned to examine whether foliar administration of irradiation chitosan (IC) and gibberellic acid (GA3) could improve the growth, structure, and biochemical characteristics of peppermint EO. Among the methods used, an absolute control using de-ionized water sprayed, an acetic acid control using 0.03% acetic acid, a mixture of both (GA3+IC). Most of the measures were significantly improved by the combination treatment when compared to the control. It results in the EO content rose to 27.9%, EO yield increased to 61.29%, menthol content increased to 15.4%, and menthone content increased to 5.85%. Nonetheless, there was a 1.46% decrease in the menthyl acetate concentration [59].

Ahmad et al.'s investigation [60] looked into the possibility of using ICN to help peppermint plants that are under arsenic (As) stress. In the experiment, 40 days after transplanting (DAT), plants received treatment using ICN (80 mg L^{-1}) either individually or when combined with As (10, 20, or 40 mg/kg of soil, as $Na_2HAsO_4 \cdot 7H_2O$). The effects on plant development, photosynthesis, and antioxidant compounds were evaluated at 150 DAT. As stress significantly reduces plant growth, affects photosynthesis, and modifies both enzymatic (superoxide dismutase and ascorbate peroxidase) and nonenzymatic (glutathione) antioxidants. At 40 mg/kg, ICN dramatically reduced the amount of peppermint EO and total phenols by 13.8% and 16.0%, respectively. It also greatly reduced the activities of phenylalanine ammonia lyase (PAL) and deoxy-D-xylulose-5-phosphate reductoisomerase (DXR) to 12.8% and 14.6%, respectively. Applying ICN improved the activities of antioxidant enzymes, promoted photosynthesis, and enhanced the accumulation of secondary metabolic products (EOs and phenols), all of which helped to mitigate the harmful effects of As toxicity in peppermint plants. Total phenols (up 17.3%) and EOs (up 36.4%) are positively correlated with ICN-stimulated increases in PAL and DXR activity (65.9% and 28.9%, respectively) relative to the control. In summary, this study showed that foliar administration of ICN (80 mg L^{-1}) successfully improved peppermint physiology and development while eliminating As-induced toxicity.

Ahmad [46] conducted a study in which different doses of irradiated sodium alginate (ISA) were used as foliar sprays and had a major impact on all growth characteristics, including shoot and root lengths and dry and fresh weights of peppermint plants. After the treatment with 90 mg L^{-1} of ISA, the root and shoot length of peppermint were increased by 32.4% and 51.6%, respectively. Moreover, dry and fresh weights were also improved by 50.9% and 44%. The foliar administration of ISA had a discernible impact on the physiological and biochemical parameters of peppermint, including whole carotenoids and chlorophyll and content, carbonic anhydrase and nitrate reductase activities, and amounts of potassium, phosphorus, and nitrogen in leaves. As compare to control, the chlorophyll and carotenoid contents were improved by 14.5% and 18.4% when treated with 90 mg L^{-1} of ISA. Carbonic anhydrase and nitrate reductase activity were also improved by 20.9% and 13.9%. The contents of leaf nitrogen, potassium, and phosphorus also showed significant improvement when treated with ISA. The topical application of ISA resulted in an increase in secondary metabolic activity, yield, and EO content of the plant. Both the EO content and the EO output of each plant showed a significant improvement with the use of the ISA. It has been noted that ISA, topically applied, modifies the content of active components. The yield was significantly impacted in comparison to the control, even though there was only a minor upsurge in the content of these monoterpenes. Menthone content had a different reaction than menthol content, which saw a favorable impact. Only at greater ISA concentrations, the amount of methyl acetate content significantly enhanced. At 90 mg L^{-1} of ISA, the contents of menthone and menthyl acetate were improved by 3.4%, 16.6%, and 21.1%, respectively.

Another investigation was carried out by Ahmad et al. [58] in which Fourier transform infrared, ultraviolet–visible (UV–vis), and nuclear magnetic resonance (NMR) spectroscopic investigation were used to characterize the RPP before and after irradiation; an attempt to acquire understandings into the potential rapport among the growth-promoting action, structural alteration, and molecular weight diminution mediated by radiation. Furthermore, different concentrations of carrageenan treated with radiation (IC) were applied topically to assess the performance of *M.*

piperita L. Out of the several IC dosage [0 (control), un-irradiated carrageenan (UC), 40, 80, 120, 160, and 200 mg L_{-1}], 80 mg L_{-1}, IC was found to have the greatest positive effect for the maximum of the parameters examined. In contrast to the control and UC, the application of 80 mg L_{-1} IC resulted in a 65.9% and 35.6% increase in the activity of rubisco and PAL, respectively. The maximum augmentation in the quantity (32.8%) and yield (88.3%) of EO was achieved by applying 80 mg L_{-1} IC. The gas chromatography results showed that applying IC, as compared to the control and UC, reduced the concentrations of menthone and menthyl acetate but raised the contents of menthol and 1,8-cineole.

Jahan [74] studied the impact of numerous fractions of irradiated carrageenan on the growth physiology and yield of peppermint plants. Different fractions of (IC-F1, IC-F2, IC-F3, IC-F4, IC-F5) along with control and unirradiated carrageenan, IC-F3 significantly enhanced the growth, physiological parameters, EO content, and yield of plants. Characterization of carrageenan and its fractions was also studied by scanning electron microscope image, IR, UV–Vis, and NMR spectroscopies. Density and diameter of oil accumulating glands, that is, PGTs, were also increased by the IC-F3 treatment.

7.5 CONCLUSION

To meet the challenges of the demand for pharmaceutically important alkaloids due to the exponential growth of the population around the world, it is seriously and urgently needed to augment the production of alkaloids and EOs by growing medicinal and aromatic plants, ensuring their sustainable use and biodiversity conservation. Several different strategies have been used to tap the available potential of different plants of medicinal and healthcare importance. In the last 10 years, naturally occurring polymers in their depolymerized form have been successfully used in agriculture, as oligomers (oligosaccharides) derived from the depolymerization of natural RPPs are thought to have strong growth-promoting properties that stimulate plant development. Among the naturally occurring RPPs, chitosan, sodium alginate, and carrageenan have been used to improve plant growth and development, thereby improving the production of medicinally important compounds in several plants. To combat these ongoing problems and maximize the production of alkaloids and EOs in medicinally significant plants, more research on the application of natural RPPs is required. This is because a number of biotic and abiotic factors, as well as the depletion of environmental conditions, impede the growth and development of plants.

KEYWORDS

- **menthone**
- **scanning electron microscope**
- **peltate glandular trichomes**
- **monoterpenes**
- **carrageenan**

REFERENCES

1. Akdogan, M., Gultekin, F., Yontem, M. Effect of *Mentha piperita* (Labiatae) and *Mentha spicata* (Labiatae) on iron absorption in rats. *Toxicology and Industrial Health*, 2004, *20*(6–10), 119–122. https://doi.org/10.1191/0748233704th206oa.
2. Akdogan, M., Ozguner, M., Aydin, G., Gokalp, O. Investigation of biochemical and histopathological effects of *Mentha piperita* (Labiatae) and *Mentha spicata* (Labiatae) on liver tissue in rats. *Human and Experimental Toxicology*, 2004, *23*(1), 21–28. https://doi.org/10.1191/0960327104ht412oa.
3. Maffei, M., Sacco, T. Chemical and morphometrical comparison between two peppermint notomorphs. *Planta Medica*, 1987, *53*(2), 214–216. https://doi.org/10.1055/s-2006-962675.
4. Akdogan, M., Kilinç, I., Oncu, M., Karaoz, E., Delibas, N. Investigation of biochemical and histopathological effects of *Mentha piperita* L. and *Mentha spicata* L. on kidney tissue in rats. *Human and Experimental Toxicology*, 2003, *22*(4), 213–219. https://doi.org/10.1191/0960327103ht332oa.
5. Sastri, B.N. The wealth of India. A dictionary of indian raw materials and industrial products. Raw materials. In: *Council of Scientific and Industrial Research.* (1950); pp. 342–344.
6. Saller, R. Peppermint (*Mentha piperita*), medicinal plant of the year 2004. *Forschende Komplementärmedizin und Klassische Naturheilkunde*, 2004, 11(1), 6–7. https://doi.org/10.1159/000077189.
7. Lorenzi, H., Matos, F.J.A. *Medicinal Plants in Brazil: Native and Exotic*. Nova Press. 2002, pp. 1–512.

8. Marcum, D.B., Carlson, H.L., Klonsky, K.M., Livingston, P. *University of California-Cooperative Extension Sample Costs to Establish a Mint Stand and Produce Peppermint Oil*. University of California Cooperative Extension. 2004.
9. Lawrence, B.M. Mint: *The Genus Mentha. Medicinal and Aromatic Plants Industrial Profiles*, CRC Press. 2007.
10. NSRC. An evidence-based systematic review of peppermint (*Mentha piperita*); Natural Standard Research Collaboration; 2012. from http://www.naturalstandard.com.
11. MIRC. Mint Industry Research Council; 2017. from http://usmintindustry.org.
12. Baytop, T. *Therapy with Medicinal Plants in Turkey-Past and Present*, 2nd edn. Istanbul, Turkey. 1999, pp. 348–349.
13. Kline, R.M., Kline, J.J., Di Palma, J., Barbero, G.J. Enteric-coated, pH-dependent peppermint oil capsules for the treatment of irritable bowel syndrome in children. *The Journal of Pediatrics*, 2019, *138*(1), 125–128. https://doi.org/10.1067/mpd.2001.109606.
14. Shaikh, S., Yaacob, H.B., Rahim, Z.H.A. Potential benefits as a safe insecticide and natural food preservative of mint (*Mentha* spp.) a review. *Asian Journal of Biomedical and Pharmaceutical Sciences*, 2014, *4*, 1–12.
15. Loolaie, M., Moasef, N., Rasouli, H., Adibi, H. Peppermint and its functionality: A review. *Archives of Clinical Microbiology*, 2017, *8*, 1–16.
16. Mohsenzadeh, M. Evaluation of antibacterial activity of selected Iranian essential oils against *Staphylococcus aureus* and *Escherichia coli* in nutrient broth medium. *Pakistan Journal of Biological Sciences*, 2007, *10*(20), 3693–3697. https://doi.org/10.3923/pjbs.2007.3693.3697.
17. Baliga, M.S., Rao, S. Radio-protective potential of mint: A brief review. *Journal of Cancer Research and Therapeutics*, 2010, *6*(3), 255–262. https://doi.org/10.4103/0973-1482.73336.
18. OECD. Menthol. In: *OECD Environment, Health and Safety Publications*. Organization for Economic Co-operation and Development: Paris, France. 2012; pp. 72–74.
19. Singh, R., Shushni, M.A.M., Belkheir, A. Antibacterial and antioxidant activities of *Mentha piperita* L. *Arabian Journal of Chemistry*, 2015, *8*(3), 322–328. https://doi.org/10.1016/j.arabjc.2011.01.019.
20. Brun, N., Colson, M., Perrin, A., Voirin, B. Chemical and morphological studies of the effects of aging on monoterpene composition in *Mentha piperita* Leaves. *Canadian Journal of Botany*, 1991, *69*(10), 2271–2278. https://doi.org/10.1139/b91-285.
21. Turner, G.W., Gershenzon, J., Croteau, R.B. Development of peltate glandular trichomes of peppermint. *Plant Physiology*, 2000, *12*, 665–680.
22. McCaskill, D., Croteau, R. Monoterpene and sesquiterpene biosynthesis in glandular trichomes of peppermint (*Mentha piperita*) rely exclusively on plastid derived isopentenyl diphosphate. *Planta*, 1995, *197*(1), 49–56. https://doi.org/10.1007/BF00239938.
23. Neeraj, T., Prakash, A., Seema, Y. Antimicrobial activity and medicinal values of essential oil of *Mentha piperita* L. *International Journal of Engineering and Innovative Technology*, 2013, *2*, 214–248.

24. Alankar, S. A review on peppermint oil. *Asian Journal of Pharmaceutical and Clinical Research*, 2009, *2*, 27–33.
25. Riachi, L.G., De Maria, C.A. Peppermint antioxidants revisited. *Food Chemistry*, 2015, *176*, 72–81. https://doi.org/10.1016/j.foodchem.2014.12.028.
26. Salehi, B., Stojanović-Radić, Z., Matejić, J., Sharopov, F., Antolak, H., Kręgiel, D., Sen, S., Sharifi-Rad, M., Acharya, K., Sharifi-Rad, R., Martorell, M., Sureda, A., Martins, N., Sharifi-Rad, J. Plants of genus Mentha: From farm to food factory. *Plants (Basel)*, 2018, *7*(3), 70. https://doi.org/10.3390/plants7030070.
27. Hendriks, H. Pharmaceutical aspects of some mentha herbs and their essential oils. *Perfum Flavor*, 1998, *23*, 15–23.
28. Mandal, S., Džudžević-Čančar, H., Dedić, A., Alispahić, A., Huremović, J. Mineral and metal content in wild and cultivated mentha species from Bosnia and Herzegovina. In: *6th World Congress on Medicinal and Aromatic Plants, WOCMAP VI Famagusta*. Northern Cyprus. 2019.
29. Gershenzon, J., McConkey, M.E., Croteau, R.B. Regulation of monoterpene accumulation in leaves of peppermint. *Plant Physiology*, 2000, *122*(1), 205–214. https://doi.org/10.1104/pp.122.1.205.
30. Wise, M.L., Croteau, R.O. Monoterpene biosynthesis. *Comprehensive Natural Products Chemistry*, 1999, *2*, 97–153.
31. Hallahan, D.L. Monoterpenoid biosynthesis in glandular trichomes of labiatae plants. *Advances in Botanical Research*, 2000, *31*, 77–120. https://doi.org/10.1016/S0065-2296(00)31007-2.
32. Courtois, J. Oligosaccharides from land plants and algae: Production and applications in therapeutics and biotechnology. *Current Opinion in Microbiology*, 2009, *12*(3), 261–273. https://doi.org/10.1016/j.mib.2009.04.007.
33. De Jesus Raposo, M.F., de Morais, A.M.B., de Morais, R.M.S.C. Marine polysaccharides from algae with potential biomedical applications. *Marine Drugs*, 2015, *13*(5), 2967–3028. https://doi.org/10.3390/md13052967.
34. Raposo, M.F., de Morais, R.M., Bernardo de Morais, A.M. Bioactivity and applications of sulphated polysaccharides from marine microalgae. *Marine Drugs*, 2013, *11*(1), 233–252. https://doi.org/10.3390/md11010233.
35. Ahmadi, A., Zorofchian Moghadamtousi, S.Z., Abubakar, S., Zandi, K. Antiviral potential of algae polysaccharides isolated from marine sources: A review. *BioMed Research International*, 2015, *2015*, 825203. https://doi.org/10.1155/2015/825203.
36. Russell, N.J., Nichols, D.S. Polyunsaturated fatty acids in marine bacteria—A dogma rewritten. *Microbiology (Reading)*, 1999, *145*(4), 767–779. https://doi.org/10.1099/13500872-145-4-767.
37. Rocha de Souza, M.C., Marques, C.T., Guerra Dore, C.M., Ferreira da Silva, F.R., Oliveira Rocha, H.A., Leite, E.L. Antioxidant activities of sulfated polysaccharides from brown and red seaweeds. *Journal of Applied Phycology*, 2007, *19*(2), 153–160. https://doi.org/10.1007/s10811-006-9121-z.
38. Campo, V.L., Kawano, D.F., da Silva Jr, D.B., Carvalho, I. Carrageenans: Biological properties, chemical modifications and structural analysis-a review. *Carbohydrate Polymers*, 2009, *77*(2), 167–180. https://doi.org/10.1016/j.carbpol.2009.01.020.

39. Shahidi, F., Ambigaipalan, P. Novel functional food ingredients from marine sources. *Current Opinion in Food Science*, 2015, *2*, 123–129. https://doi.org/10.1016/j.cofs.2014.12.009.
40. Lee, Y.-E., Kim, H., Seo, C., Park, T., Lee, K.B., Yoo, S.-Y., Hong, S.-C., Kim, J.T., Lee, J. Marine polysaccharides: Therapeutic efficacy and biomedical applications. *Archives of Pharmacal Research*, 2017, *40*(9), 1006–1020. https://doi.org/10.1007/s12272-017-0958-2.
41. Hien, N.Q., Luan, L.Q., Hanh, T.T., Hai, L. *Radiation Processing Technology for Production of Plant Growth Promoter from Brown Seaweed and Plant Protector from Shrimp Shell; Nuclear Research Institute*. IAEA. SM-365/23. 2009; p. 45.
42. De la Rosa, A.M., Abad, L.V., Relleve, L., Aranilla, C. Radiation-modified carrageenan for agricultural and health care applications. *IAEA Report, International Atomic Energy Agency (IAEA) Project Coordination Meeting on Radiation Processing of Chitin/Chitosan*, Bangkok, Thailand, March 18–20, 2002.
43. Abad, L.V., Aurigue, F.B., Relleve, L.S., Montefalcon, D.R.V., Lopez, G.E.P. Characterization of low molecular weight fragments from gamma irradiated κ-carrageenan used as plant growth promoter. *Radiation Physics and Chemistry*, 2016, *118*, 75–80. https://doi.org/10.1016/j.radphyschem.2015.03.001.
44. Abad, L.V., Dean, G.F.O., Magsino, G.L., Dela Cruz, R.M.M., Tecson, M.G., Abella, M.E.S., Hizon, M.G.S. Semi-commercial scale production of carrageenan plant growth promoter by e-beam technology. *Radiation Physics and Chemistry*, 2018, *143*, 53–58. https://doi.org/10.1016/j.radphyschem.2017.07.009.
45. Abad, L.V., Aurigue, F.B., Montefalcon, D.R.V., Manguiat, P.H., Carandang, F.F., Mabborang, S.A., Hizon, M.G.S., Abella, M.E.S. Effect of radiation-modified kappa-carrageenan as plant growth promoter on peanut (Arachis hypogaea L.). *Radiation Physics and Chemistry, 2018*, 153, 239–244. https://doi.org/10.1016/j.radphyschem.2018.10.005.
46. Ahmad, B. Influence of silicon nanoparticles and radiation-processed chitosan and sodium alginate on growth, yield and essential oil quality of peppermint (*Mentha piperita* L.). Ph.D. Thesis; Aligarh Muslim University. 2017.
47. Singh, M., Khan, M.M.A., Uddin, M., Naeem, M., Qureshi, M.I. Proliferating effect of radiolytically depolymerized carrageenan on physiological attributes, plant water relation parameters, essential oil production and active constituents of Cymbopogon flexuosus under drought stress. *PLoS One*, 2017, *12*(7), e0180129. https://doi.org/10.1371/journal.pone.0180129.
48. Ahmed, K.B.M., Khan, M.M.A., Siddiqui, H., Jahan, A. Chitosan and its oligosaccharides, a promising option for sustainable crop production—A review. *Carbohydrate Polymers*, 2019, *17*, 115–331.
49. Ali, A., Khan, M.M.A., Uddin, M., Naeem, M., Idrees, M., Hashmi, N., Dar, T.A., Varshney, L. Radiolytically depolymerized sodium alginate improves physiological activities, yield attributes and composition of essential oil of *Eucalyptus citriodora* Hook. *Carbohydrate Polymers*, 2014, *112*, 134–144. https://doi.org/10.1016/j.carbpol.2014.05.070.
50. Dar, T.A., Uddin, M., Khan, M.M.A., Ali, A., Mir, S.R., Varshney, L. Effect of Co-60 gamma irradiated chitosan and phosphorus fertilizer on growth, yield and Trigonelline

content of Trigonella foenum-graecum L. *Journal of Radiation Research and Applied Sciences, 2015*, 8(3), 446–458. https://doi.org/10.1016/j.jrras.2015.03.008.
51. Dar, T.A., Uddin, M., Khan, M.M.A., Ali, A., Varshney, L. Modulation of alkaloid content, growth and productivity of Trigonella foenum-Graecum L. Using irradiated sodium alginate in combination with soil applied phosphorus. *Journal of Applied Research on Medicinal and Aromatic Plants*, 2016, *3*(4), 200–210. https://doi.org/10.1016/j.jarmap.2016.05.003.
52. Shabbir, A., Khan, M.M.A., Sadiq, Y., Jaleel, H., Ahmad, B., Uddin, M. Regulation of functional activities and essential oil production in Vetiveria zizanioides L. Nash after γ-irradiated sodium alginate elicitation. *Turkish Journal of Biology*, 2017, *41*, 661–672. https://doi.org/10.3906/biy-1704-6.
53. Sadiq, Y., Khan, M.M.A., Shabbir, A., Ahmad, B., Jaleel, H., Uddin, M., Varshney, L. Structural re-arrangement of depolymerized sodium alginate enriches peltate glandular trichomes and essential oil production of spearmint. *International Journal of Biological Macromolecules*, 2017, *105*(1), 1043–1050. https://doi.org/10.1016/j.ijbiomac.2017.07.134.
54. Jaleel, H., Khan, M.M.A., Ahmad, B., Shabbir, A., Sadiq, Y., Uddin, M., Varshney, L. Elucidating the dynamics of physiology and essential oil production in lemongrass using alginate oligomers under field conditions. *Journal of Functional and Environmental Botany*, 2017, *7*(1), 26–35. https://doi.org/10.5958/2231-1750.2017.00004.X.
55. Jaleel, H., Khan, M.M.A., Ahmad, B., Shabbir, A., Sadiq, Y., Uddin, M., Varshney, L. Essential oil and citral production in field-grown lemongrass in response to gamma-irradiated chitosan. *Journal of Herbs Spices & Medicinal Plants*, 2017, *10*, 1–5.
56. Naeem, M., Nabi, A., Aftab, T., Khan, M.M.A. Oligomers of carrageenan regulate functional activities and artemisinin production in Artemisia annua L. Exposed to arsenic stress. *Protoplasma*, 2020, *257*(3), 871–887. https://doi.org/10.1007/s00709-019-01475-y.
57. Ahmad, B., Khan, M.M.A., Jaleel, H., Sadiq, Y., Shabbir, A., Uddin, M. Exogenously sourced γ-irradiated chitosan-mediated regulation of growth, physiology, quality attributes and yield in *Mentha piperita* L. *Turkish Journal of Biology*, 2017, *41*, 388–401. https://doi.org/10.3906/biy-1608-64.
58. Ahmad, B., Jahan, A., Sadiq, Y., Shabbir, A., Jaleel, H., Khan, M.M.A. Radiation-mediated molecular weight reduction and structural modification in carrageenan potentiates improved photosynthesis and secondary metabolism in peppermint (*Mentha piperita* L.). *International Journal of Biological Macromolecules*, 2019, *124*, 1069–1079. https://doi.org/10.1016/j.ijbiomac.2018.12.022.
59. Ahmad, B., Jaleel, H., Shabbir, A., Khan, M.M.A., Sadiq, Y. Concomitant application of depolymerized chitosan and GA3 modulates photosynthesis, essential oil and menthol production in peppermint (*Mentha piperita* L.). *Scientia Horticulturae*, 2019, *246*, 371–379. https://doi.org/10.1016/j.scienta.2018.10.031.
60. Ahmad, B., Dar, T.A., Khan, M.M.A., Ahmad, A., Rinklebe, J., Chen, Y., Ahmad, P. Oligochitosan fortifies antioxidative and photosynthetic metabolism and enhances secondary metabolite accumulation in arsenic-stressed peppermint. *Frontiers in Plant Science*, 2022, *13*, 987746. https://doi.org/10.3389/fpls.2022.987746.
61. Ahmed, K.B.M., Khan, M.M.A., Jahan, A., Siddiqui, H., Uddin, M. Gamma Rays induced acquisition of structural modification in chitosan boosts photosynthetic machinery,

enzymatic activities and essential oil production in citronella grass (Cymbopogon Winterianus Jowitt). *International Journal of Biological Macromolecules*, 2020, *145*, 372–389. https://doi.org/10.1016/j.ijbiomac.2019.12.130.

62. Therkelsen, G.H. In: Whistler, R.L., BeMiller, J.N. (Eds.) *Industrial Gums: Polysaccharides and Their Derivatives*. Academic Press, 1993; pp. 145–180.
63. Van de Velde, F., Knutsen, S.H., Usov, A.I., Rollema, H.S., Cerezo, A.S. 1H and 13C high resolution NMR spectroscopy of carrageenans: Application in research and industry. *Trends in Food Science & Technology*, 2002, *13*(3), 73–92. https://doi.org/10.1016/S0924-2244(02)00066-3.
64. Van de Velde, F., Pereira, L., Rollema, H.S. The revised NMR chemical shift data of carrageenans. *Carbohydrate Research*, 2004, *339*(13), 2309–2313. https://doi.org/10.1016/j.carres.2004.07.015.
65. Stanley, N.F. Production and Utilization of Products from Commercial Seaweeds. In F.A.O. Fish. Tech. Pap. McHugh, D.J. (Eds.), *Food and Agriculture Organization*. 1987; pp. 116–146.
66. Stanley, N.F. Carrageenans. In: Harris P. (Eds.), *Food Gels*. Elsevier Science Publishers Ltd. 1990; p. 96.
67. Mollion, J. Exploitation of Algal Carrageenans: Global Situation and Perspectives in Senegal. In: *Marine Chemistry for Development*. Kornprobst, J.M., (Eds.). 1983, pp. 84–95.
68. Parekh, R.G., Rao, A.V., Chauhan, V.D. Carrageenan from marine red alga, hypnea valentiae (Turn.) Mont. *Phykos*, 1988, *26*, 146–151.
69. Campo, V.L., Kawano, D.F., da Silva Jr, D.B., Carvalho, I. Carrageenans: Biological properties, chemical modifications and structural analysis—A review. *Carbohydrate Polymers*, 2009, *77*(2), 167–180. https://doi.org/10.1016/j.carbpol.2009.01.020
70. Prajapati, V.D., Maheriya, P.M., Jani, G.K., Solanki, H.K. Carrageenan: A natural seaweed polysaccharide and its applications. *Carbohydrate Polymers*, 2014, *105*, 97–112. https://doi.org/10.1016/j.carbpol.2014.01.067.
71. Abad, L.V., Kudo, H., Saiki, S., Nagasawa, N., Tamada, M., Katsumura, Y., Aranilla, C.T., Relleve, L.S., De La Rosa, A.M. Radiation degradation studies of carrageenans. *Carbohydrate Polymers*, 2009, *78*(1), 100–106. https://doi.org/10.1016/j.carbpol.2009.04.009.
72. Abad, L.V., Saiki, S., Nagasawa, N., Kudo, H., Katsumura, Y., De La Rosa, A.M. NMR Analysis of fractionated irradiated k-carrageenan oligomers as plant growth promoter. *Radiation Physics and Chemistry*, 2011, *80*(9), 977–982. https://doi.org/10.1016/j.radphyschem.2011.04.006.
73. Jiang, Y.P., Guo, X.K., Tian, X.F. Synthesis and NMR structural analysis of O-succinyl derivative of low-molecular-weight-k-carrageenan. *Carbohydrate Polymers*, 2005, *61*(4), 399–406. https://doi.org/10.1016/j.carbpol.2005.05.016.
74. Jahan, A. Influence of Different Fractions of Radiation-Processed Carrageenan on Performance of Some Mentha species. *Ph.D. Thesis*; Aligarh Muslim University, 2020.
75. Peniston, Q.P., Johnson, E.L. Process for the manufacture of chitosan. *U.S. Patent 4, 195175. Roberts GAF (1992) Chitin Chemistry*. 1980; pp. 54–83.
76. Isik, M., Sardon, H., Mecerreyes, D. Ionic liquids and cellulose: Dissolution, chemical modification and preparation of new cellulosic materials. *International*

Journal of Molecular Sciences, 2014, *15*(7), 11922–11940. https://doi.org/10.3390/ijms150711922.

77. Tikhonov, V.E., Stepnova, E.A., Babak, V.G., Yamskov, I.A., Palma-Guerrero, J., Jansson, H.B., Lopez-Llorca, L.V., Salinas, J., Gerasimenko, D.V., Avdienko, I.D., Varlamov, V.P. Bactericidal and antifungal activities of a low molecular weight chitosan and its N-/2(3)-(dodec-2-enyl)succinoyl/-derivatives. *Carbohydrate Polymers*, 2006, *64*(1), 66–72. https://doi.org/10.1016/j.carbpol.2005.10.021.
78. Wojdyla, A.T., Orlikowski, L.B., Struszczyk, H. Chitosan for the control of leaf pathogens. *Chitin Enzymol.* 2001, 191–196.
79. Muzzarelli, R.A., Mattioli-Belmonte, M., Tietz, C., Biagini, R., Ferioli, G., Brunelli, M.A., Fini, M., Giardino, R., Ilari, P., Biagini, G. Stimulatory effect on bone formation exerted by a modified chitosan. *Biomaterials*, 1994, *15*(13), 1075–1081. https://doi.org/10.1016/0142-9612(94)90093-0.
80. Hien, N.Q. Radiation degradation of chitosan and some biological effects. Radiation processing of polysaccharides. *International Atomic Energy Agency Austria*, 2004, 1422, 67–73.
81. Haji-Saeid, M., Safrany, A., Sampa, M.H.D.O., Ramamoorthy, N. Radiation processing of natural polymers: The IAEA contribution. *Radiation Physics and Chemistry*, 2010, *79*(3), 255–260. https://doi.org/10.1016/j.radphyschem.2009.11.001.
82. Abbasi, N.A., Iqbal, Z., Maqbool, M., Hafiz, I.A. Postharvest quality of mango (Mangifera indica L.) fruit as affected by chitosan coating. *Pakistan Journal of Botany*, 2009, *41*, 343–357.
83. Islam, A., Yasin, T. Controlled delivery of drug from pH sensitive chitosan/poly (Vinyl Alcohol) blend. *Carbohydrate Polymers*, 2012, *88*(3), 1055–1060. https://doi.org/10.1016/j.carbpol.2012.01.070.
84. Ravi Kumar, M.N.V.R. A review on chitin and chitosan applications. *Reactive and Functional Polymers*, 2000, *46*(1), 1–27. https://doi.org/10.1016/S1381-5148(00)00038-9.
85. Photchanachai, S., Singkaew, J., Thamthong, J. Effects of chitosan seed treatment on Colletotrichum sp. and seedling growth of chili cv. 'Jinda'. *Acta Horticulturae*, 2006, *712*, 585–590. https://doi.org/10.17660/ActaHortic.2006.712.70.
86. Hadwiger, L.A., McBride, P.O. Low-level copper plus chitosan applications provide protection against late blight of potato. *Plant Health Progress*, 2006, *7*(1), 150–168. https://doi.org/10.1094/PHP-2006-0406-01-RS
87. Nge, K.L., Nwe, N., Chandrkrachang, S., Stevens, W.F. Chitosan as a growth stimulator in orchid tissue culture. *Plant Science*, 2006, *170*(6), 1185–1190. https://doi.org/10.1016/j.plantsci.2006.02.006.
88. Vasyukova, N.I., Zinov'Eva, S.V., Il'inskaya, L.I., Perekhod, E.A., Chalenko, G.I., Gerasimova, N.G., Il'ina, A.V., Varlamov, V.P., Ozeretskovskaya, O.L. Modulation of plant resistance to diseases by water-soluble chitosan. *Applied Biochemistry and Microbiology*, 2001, *37*(1), 103–109. https://doi.org/10.1023/A:1002865029994.
89. Kuchitsu, K., Kosaka, H., Shiga, T., Shibuya, N. EPR evidence for generation of hydroxyl radical triggered by N-acetyl-chito-oligosaccharide elicitor and a protein-phosphatase inhibitor in suspension-cultured rice cells. *Protoplasma*, 1995, *188*(1–2), 138–142. https://doi.org/10.1007/BF01276805.

90. Nojiri, H., Sugimori, M., Yamane, H., Nishimura, Y., Yamada, A., Shibuya, N., Kodama, O., Murofushi, N., Omori, T. Involvement of jasmonic acid in elicitor-induced phytoalexin production in suspension-cultured rice cells. *Plant Physiology*, 1996, *110*(2), 387–392. https://doi.org/10.1104/pp.110.2.387.
91. Takai, K., Ohtsuka, T., Senda, Y., Nakao, M., Yamamoto, K., Matsuoka, J., Hirai, Y. Antibacterial properties of antimicrobial-finished textile products. *Microbiology and Immunology*, 2002, *46*(2), 75–81. https://doi.org/10.1111/j.1348-0421.2002.tb02661.x.
92. Xu, X.Q., Shen, H., Xu, J.R., Xie, M.Q., Li, X.J. The colloidal stability and core-shell structure of magnetite nanoparticles coated with alginate. *Applied Surface Science*, 2006, *253*(4), 2158–2164. https://doi.org/10.1016/j.apsusc.2006.04.015.
93. Pourhadi, M., Badi, H.N., Mehrafarin, A., Omidi, H., Hajiaghaee, R. Phytochemical and growth responses *Mentha piperita* to foliar application of biostimulants under greenhouse and field conditions. *Herba Polonica*, 2018, *64*(2), 1–12. https://doi.org/10.2478/hepo-2018-0010.

CHAPTER 8

Gamma Irradiation-Mediated Restructuring of Sodium Alginate Enhances the Content of Essential Oil in Peppermint by Improving Its Growth, Nutrient Content, and Enzymatic Activities

BILAL AHMAD,[1*] TARIQ AHMAD DAR,[2] AJMAT JAHAN,[3] MUDASIR IRFAN DAR,[4] SUHAIL BASHIR,[5], and M. MASROOR A. KHAN[6]

[1]*Department of Botany, Government Degree college Khansahib Budgam, University of Kashmir*

[2]*Department of Botany, Sri Pratap College, Srinagar, Cluster University Srinagar, India*

[3]*Department of Botany, Aligarh Muslim University, Aligarh, Uttar Pradesh, India*

[4]*Department of Botany, Government Degree College for Women, Kupwara, University of Kashmir, India*

[5]*Department of Botany, Government Degree College Sogam, Kupwara, University of Kashmir, India*

[6]*Advanced Plant Physiology Section, Department of Botany, Aligarh Muslim University, Aligarh, Uttar Pradesh, India*

*Corresponding author

Peppermint: Novel Strategies for Performance Improvement. Bilal Amhad, Tariq Ahmad Dar, and Khalid Rehman Hakeem (Eds.)

DOI: 10.1201/9781779641557-8

ABSTRACT

Restructuring of biopolymers using gamma-irradiation and employing them to enrich the content of secondary products of pharmaceutical significance has been the focus of the research during the last decade, in view of the abundance of biopolymers in nature and their eco-friendly nature. Irradiation, in addition to polymer-restructuring, reduces the molecular weight and yields low molecular weight oligomers, consequently resulting in the attainment of growth promotion activity. In the current study, sodium alginate after gamma-radiation mediated re-structuring was employed in the form of graded concentrations to test whether the content and productivity of peppermint would be boosted. Interestingly, the growth, content of light capturing pigments (total chlorophyll and total carotenoids), enzyme activities carbonic anhydrase, nitrate reductase, content of critical essential macronutrients (NPK), percentage and yield of essential oil and monoterpene (menthol) were enhanced significantly by employing 90 mg L^{-1} concentration of restructured sodium alginate in the form of aqueous formulations. The content of menthone was, however, reduced by this concentration. Conclusively, restructured sodium alginate may be employed to improve the performance of medicinally imperative plants like peppermint.

8.1 INTRODUCTION

Polysaccharides, being the most copious biological resources on the earth, contribute to their widespread application in industry and agriculture [9, 14]. Their agricultural usage is still quite modest considering their immense potential and versatility, even though it has significantly escalated recently. Polysaccharide re-structuring and molecular weight diminution, an advantageous outcome of gamma-irradiation, has developed a considerable interest among agricultural scientists, and the research studies of the last decade have elucidated that such processed biopolymers act as phytoelicitors [1, 10]. Alginate is one among the most plentiful polysaccharides extracted from molluscs and phaeophycean algae with extensive applications in industry and agriculture [37, 39]. Sodium alginate, restructured by varied dose rate of gamma-radiations, attains incomparable phytoelicitor activity which is endorsed to decreased molecular weight, polymer re-structuring and creation of new functional-groups not present in the unprocessed form [38, 41, 43]. Aqueous application of such restructured oligomers through foliage perks up the plant metabolism by ameliorating nutrient assimilation, enzymatic up-regulation, and photosynthate accumulation and diversion toward secondary metabolism. Restructured-oligosaccharides up-regulate the activities of principle metabolic enzymes, notably carbonic anhydrase (CA), deoxy-D-xylulose-5-phosphate reducto-isomerase, L-tryptophan decarboxylase, nitrate reductase (NR), phosphoenolpyruvate carboxylase (PepCo), RuBisCo, phenylalanine-ammonia lyase, consequently boosting the plant performance [6, 29, 30, 42]. Ionizing radiations constitute the most capable and resourceful method of polymer modification, being a single-step restructuring procedure. Irradiation-induced re-structuring splits up the glycosidic bonds and creates unsaturation without altering the chemical structure of the biopolymer considerably [7, 8, 13, 20, 21, 40]. Additionally, altered cross-linkages have been monitored to introduce new functional groups like carbonyl-groups in addition to the unsaturation, which is endorsed to the growth promotion activity. Such restructuring facilitates ligand-receptor binding, which consequently assists in effective gene expression by stimulating diverse signaling pathways like G-protein and mitogen-activating protein kinase pathways [5]. Furthermore, the eco-friendly nature bestows the biopolymers, an edge over the conventional phytoregulators being nontoxic and biodegradable [5, 6, 25].

Plants constitute a treasure of medicine and are employed as traditional and Ayurvedic medicine to treat a range of maladies. The medicinal significance of these plants is because of bioactive compounds of secondary origin, notably terpenoids, phenolics, and alkaloids. *Mentha piperita* (peppermint), a plant with enormous pharmaceutical implications, notably carminative, serotonergic, antispasmodic, antibacterial, antifungal, antiviral, and anti-inflammatory, is mainly cultivated for its precious essential oil (EO) [26, 33, 44]. Many other properties like hepatoprotective, cytoprotective, and antiulcer, flatulence, gastritis, colitis, increased hormonal emission, vasodilation, and stimulation of bile production, augment its therapeutic implication [26, 33, 36, 46]. Peppermint, cultivated for its economically vital EO, mostly contains monoterpenes of volatile nature, among which the most prevailing constituent is menthol (up to 50%), which is recognized as one among the most preferred flavors across the globe [26, 46]. The monoterpenes (menthol, menthyl acetate, menthofuran) and polyphenols compose a dominant proportion of the EO [28, 35, 36]. Polyphenols have antioxidative properties and potentiate the plants to fight stresses of diverse nature [27, 32]. However, the content of EO in the plants is inadequate (up to 1%), and the yield is further curtailed due to a variety of reasons. The need of the hour is to adopt novel and sustainable strategies to enhance the content of EO in plants, as the backbone of the traditional drug-based pharmaceutical companies relies upon the secondary products like EOs. The present inquest was aimed at studying the influence of varied concentrations of restructured sodium alginate on the performance of peppermint.

8.2 MATERIALS AND METHOD

8.2.1 EXPERIMENTAL PLAN

This study (simple randomized design) was conducted to examine the effect of graded concentrations of radiolytically-degraded sodium alginate (ISA), namely, 30, 60, 90, and 120 mg L^{-1}, on *M. piperita* L. The experiment was carried out in the net-house of the Botany Department, Aligarh Muslim University, Aligarh, Uttar Pradesh (78°51′ E longitude, 27°52′ N latitude, 187.45 m altitude) under natural conditions. The average temperature during the experimental period was recorded to be 7.4°C (December), 7.1°C (January), 11.7°C (February), 19.3°C (March), 25.1°C (April), and 27.6°C (May). The mean precipitation was 5.1–15.3 mm, and the relative humidity was 35–74%. The plants were raised from healthy rhizomes acquired from CIMAP, Lucknow. The rhizomes, after surface sterilization in 0.02% $HgCl_2$ solution with vigorous shaking for 3 min, were thoroughly washed using deionized water (DIW). The experiment was conducted in plastic pots of 8.5 kg capacity (40 cm × 45 cm), each of which was filled with a homogenous mixture of soil and organic manure (5:1). The soil was analyzed at the Soil-Testing Laboratory, Indian Agricultural Research Institute, New Delhi, physicochemical characteristics. The soil characteristics as revealed from the report were as follows: texture-sandy loam, pH (1:2) 7.83, EC (1:2) 0.34 m mhos cm^{-1}, available N, P, and K of 142.2, 78.3, and 157.3 mg kg^{-1} of soil, respectively. During the growing season, irradiated sodium alginate (ISA) was applied through foliage on 45-day-old seedlings at an interval of 10 days by means of a hand sprayer at a seven-day interval, and the treatment was repeated seven times. DIW was used as the Control. The attributes studied during the current study were analyzed at 150 days after transplanting (DAP). A uniform basal split-dose of NPK ($N_{30+30}P_{30+30}K_{30+30}$ mg kg^{-1}) was added to the soil. Application of N, P, and K was carried out as urea, single superphosphate, and muriate of potash. The first as well as the second half were applied at the rate of 30 mg kg^{-1} of soil with a gap of one month in-between.

8.2.2 IRRADIATION OF SODIUM ALGINATE BY CO-60 GAMMA RAYS

Sodium alginate was procured from Sigma-Aldrich, New Delhi. Irradiation of the purchased polysaccharide at a dose rate of 520 kGy, was carried out using Co-60 as the source, at ISOMED Group, Bhabha Atomic Research Centre.

8.2.3 ASSESSMENT OF GROWTH ATTRIBUTES

Assessment of plant growth (shoot length and root length, and fresh and dry weights) was measured at 150 DAP. From each treatment pot, three plants were dug out and the sticked soil and foreign particles were cautiously washed using tap water. Blotting was done using blotting paper in order to eliminate the adhered water, and the length of the plant (shoot and root, separately) was computed using a measuring scale. After documenting the plant fresh weight, the plants were dried overnight at a high temperature (80°C) inside a hot air oven for 24 h, in order to record the dry weight.

8.2.4 DETERMINATION OF PHYSIOLOGICAL ACTIVITIES

8.2.4.1 TOTAL CHLOROPHYLL AND CAROTENOID CONTENT

Contents (concentration) of chlorophyll were approximated using fresh peppermint leaves according to the method as described by Lichtenthaler and Buschmann [23]. Fresh interveinal leaf tissue was ground with the help of a mortar and pestle with acetone (100%). The pigment solution obtained after purification was used to record the optical density at 470, 645, and 662 nm for the assessment of the content of chlorophylls (a and b) as well as total carotenoids, respectively, by means of a spectrophotometer (Shimadzu UV-1700, Tokyo, Japan). Total content of the pigment (chlorophyll) was determined by totaling the contents of a and b-type chlorophylls. Thereafter, the content of the assessed pigments was expressed as mg g^{-1} leaf FW.

8.2.4.2 DETERMINATION OF ACTIVITIES OF NITRATE REDUCTASE (NR) AND CARBONIC ANHYDRASE (CA)

Intact leaf-tissue assay devised by Jaworski [22] was adopted to estimate NR (EC 1.7.1.1) activity. The quantity of nitrite produced was analyzed spectrophotometrically. To a plastic vial were added: a phosphate buffer (2.5 mL, pH 7.5), potassium nitrate (KNO_3) solution (0.5 mL), and isopropanol (5%, 2.5 mL), and then fresh thinly chopped leaves (0.2 g) were put into the vial. Incubation of the vial was carried out for a duration of 2 h at 30°C temperature and thereafter, 0.4 mL content from the vial was relocated to a test tube. Subsequent to this, 0.3 mL of sulfanilamide (1%) and the same volume of *N*-(1-naphthyl) ethylenediamine dihydrochloride (0.02%) were added. Lastly, these vials were left untouched at room temperature for 20 min so as to allow the highest color development, and the content in the vial was eventually increased to a volume of 5 mL by diluting it with distilled water. The OD was noted down at 540 nm by means of the spectrophotometer, and the enzyme activity was expressed as nM NO_2^{-1} g^{-1} FW h^{-1}.

CA activity was determined by the method of Dwivedi and Randhawa [15]. Fresh leaf pieces very gauzily chopped (0.2 g) from each treatment, after transferring to a petriplate, were dipped into a 0.2 M solution of cysteine hydrochloride (10 mL) having a temperature of 4°C for 20 min duration. To each test tube was then added 4 mL of 0.2 M $NaHCO_3$ solution and 0.2 mL of 0.002% bromothymol blue. The reaction mixture in the test tubes was titrated against HCl (0.05 N), and methyl red acted as an indicator. The activity of the CA enzyme was finally evaluated as M CO_2 kg^{-1} leaf FW s^{-1}.

8.2.4.3 ESTIMATION OF NITROGEN CONTENT

Leaf N content was approximated as per Lindner [24], with minor modification by Novozamsky et al. [34]. The oven-dried leaves were powdered and digested using H_2SO_4 in digestion-assembly tubes with the help of a digestion assembly. An aliquot (10 mL) of peroxide-digested powdered-foliage material was relocated into a 50 mL volumetric flask. 2 mL of NaOH solution (2.5 N) and 1 mL of sodium silicate solution (10%) were included, respectively in order to reduce the effect of the surplus acid and circumvent turbidity. Thereafter, a graduated tube of 10 mL capacity was half-filled with the aliquot, and 0.5 mL of Nessler's reagent were supplemented to it. The OD was eventually recorded at 525 nm for estimating the content of leaf N.

8.2.4.4 ESTIMATION OF PHOSPHORUS CONTENT

The procedure devised by Fiske and Subba Row (1925), with small alterations by Rorison et al. (1993), was used to assess the leaf-P content in the peroxide-digested foliage material. A 5 mL aliquot was transferred to a graduated tube (10 mL capacity). Thereafter, 2.5% molybdic acid (1 mL) was added, followed by subsequent supplementation of 1-amino-2-naphthol-4-sulfonic acid (0.4 mL). When the content in the tube turned blue, the volume was enhanced to 10 mL by adding DDW. The OD was ultimately recorded at 620 nm by means of a spectrophotometer.

8.2.4.5 ESTIMATION OF POTASSIUM CONTENT

The peroxide digested leaf powder was also used for the estimation of potassium content in peppermint leaves using a flame photometer (model C150, AIMIL, India) following Hald's method [18]. The digested leaf material was released through an atomizer as a fine spray, like a mist, into a chamber where it was flamed. Combustion of K results in the production of light of 767 nm wavelength (violet). This light, after passing through a proper filter, was imprinted on a photoelectric cell, which consequently caused the activation of the galvanometer.

8.2.5 YIELD AND QUALITY PARAMETERS

8.2.5.1 ISOLATION AND COMPOSITIONAL ANALYSIS OF ESSENTIAL OIL

Fresh green leaves plucked from each treatment pot were finely chopped, mixed, and were subsequently taken for the assessment of the volatile EO content according to Guenther [17]. The steam distillation procedure was employed for extracting EO from the tender green parts using Clevenger's apparatus. The EO extracted was then dried over sodium sulphate (anhydrous) and stored in sealed glass vials at a low temperature of 4°C and the composition and the contents of individual monoterpenes was assessed using gas chromatography–mass spectroscopy (GC–MS) (Shimadzu

QP-2010 equipped with thermal desorption system TD 20 having (Rtx-5) fused silica capillary column with dimensions of 30 m × 0.25 mm and 0.25 μm film thickness. Temperature of the GC–MS was programmed as: detector temperature (270°C), oven temperature (275°C), and column temperature (250°C) with a 2 min hold time @ 5°C min^{-1} and thereafter 260–280°C at 10°C min^{-1} with a last and final hold time of 15 min, using He gas as carrier. The scheduled injector temperature was 260°C, the sample size comprised of 0.2 μL, and the split ratio of finalized as 1:10. Ionization energy was set at 70 eV (EI), having a mass-scan array of 40–400 amu.

8.2.6 STATISTICAL ANALYSIS

Each treatment pot was considered as one replicate, and every treatment was repeated five times. The statistical data analysis was carried out by means of SPSS-17 statistical software (SPSS Inc., Chicago, IL, USA). Comparison of the means was done employing Duncan's multiple range test at $p < 0.05$. In order to separate the means in the data tables, the standard error was applied.

8.3 RESULTS AND DISCUSSION

8.3.1 LEAF-APPLIED SODIUM ALGINATE OLIGOMERS IMPROVE GROWTH OF PEPPERMINT

Plant growth is principally regulated by a wide array of factors (endogenous and exogenous) that modulate imperative primary and secondary metabolic phenomena. Polysaccharides processed through irradiation, a novel and relatively recent class of exogenous elicitors, enhance growth in plants by manipulating key life events [6, 30, 42]. The attainment of elicitation effect in polysaccharides like sodium alginate after irradiation can prove eco-friendly as well as economical to enhance the yield of plants of economic significance. Reduced molecular weight and restructuring of functional groups leading to structural reorganization have been associated with growth promotion activity in bio-polymers [5–7, 30, 31, 40, 42]. Among the various factors, γ-radiations are the most potent agents of molecular weight reduction used and yield fruitful results as compared to acid-base hydrolysis and enzymatic breakdown [3, 4, 30, 38, 43]. The polymer under study (sodium alginate) undergoes molecular weight reduction and structural rearrangement as has been reported previously [2, 3, 40].

The current study revealed that graded concentrations of ISA applied in the form of foliar sprays significantly influenced all the growth attributes, namely, shoot and root lengths and fresh and dry weights (Table 8.1). ISA-90 mg L^{-1} significantly enhanced the shoot and root lengths of the plant by 32.4% and 51.6%, respectively, over the Control (Table 8.1). The same treatment maximally enhanced the fresh and dry weights of peppermint by 44% and 50.9%, respectively, over the Control. A progressive increase up to 90 mg L^{-1} ISA was observed in the values of these parameters, beyond which a steep decline was observed. The elicitor effect of sodium alginate-derived oligomers obtained after irradiation is evident from the current finding of the oligosaccharide-mediated elicitation of growth on account of modulation of various metabolic processes in plants. Exogenous sourcing of such oligomers evokes the intrinsic potential of the plants, thereby augmenting plant growth analogous to conventional plant growth regulators [5, 13, 20, 21, 40]. Application of processed-sodium alginate in the form of aqueous solutions augmented growth, which is endorsed to up-regulate cyclins and protein kinases dependent upon cyclins resulting in enhanced growth by increased cell division [11]. Sodium alginate oligomers are taken up by the plant foliage readily when applied as aqueous solutions and stir up the intrinsic potential, improving peppermint growth like conventional phytohormones [11, 16]. Moreover, ameliorated nutrient assimilation and enzymatic activities put in more photosynthates, which, in turn, is endorsed to improve peppermint growth [6, 5, 40, 42].

8.3.2 PHYSIOLOGICAL AND BIOCHEMICAL PARAMETERS

8.3.2.1 TOTAL CHLOROPHYLL AND TOTAL CAROTENOIDS CONTENT

Physiological and biochemical parameters (total chlorophyll and carotenoid content, CA and NR activities, content of nitrogen, phosphorus, and potassium in leaves) of peppermint exhibited an apparent influence in response to the foliar application of sodium alginate oligomers (Table 8.2). Among the graded foliar concentrations, 90 mg L^{-1} ISA improved the contents of total chlorophyll as well as total carotenoids maximally in the fresh leaves of peppermint. The percent increase in the total chlorophyll content was 14.5% while that of content of total carotenoid by 18.4% over the Control (Table 8.2).

TABLE 8.1 Effect of Foliar Spray of Different ISA Concentrations on Growth Parameters of Peppermint*

Parameters	ISA Concentrations (mg L−1)					
	0	**USA**	**ISA-30**	**ISA-60**	**ISA-90**	**ISA-120**
Shoot length (cm)	66.90 ± 3.75^{d}	67.12 ± 3.05^{d}	73.4 ± 3.73^{cd}	79.20 ± 3.65^{bc}	88.60 ± 3.83^{a}	85.30 ± 3.55^{ab}
Root length (cm)	34.70 ± 1.15^{e}	34.82 ± 1.21^{e}	38.90 ± 1.30^{d}	43.10 ± 1.18^{c}	52.60 ± 1.50^{a}	48.10 ± 1.29^{b}
Fresh weight (g)	84.37 ± 3.27^{d}	86.37 ± 3.27^{d}	95.43 ± 3.30^{c}	108.16 ± 3.35^{b}	121.52 ± 2.91^{a}	116.62 ± 3.21^{ab}
Dry weight (g)	20.03 ± 1.14^{d}	20.15 ± 1.14^{d}	23.67 ± 1.08^{c}	25.81 ± 1.2^{bc}	30.23 ± 1.05^{a}	27.43 ± 1.21^{ab}

*Means within a row followed by the same letter(s) are not significantly different ($p \leq 0.05$).

Note: Each value represents the mean of five replicates with ± S.E.

ISA: Irradiated sodium alginate; USA: Un-irradiated sodium alginate.

TABLE 8.2 Effect of Foliar Spray of Different ISA Concentrations on Total Chlorophyll Content, Total Carotenoids Content, Carbonic Anhydrase Activity, Nitrate Reductase Activity, Nitrogen Content, Phosphorous Content, and Potassium Content of Peppermint After 150 DAP (Days After Planting)*

Parameters	ISA Concentrations (mg L−1)					
	0	**USA**	**ISA-30**	**ISA-60**	**ISA-90**	**ISA-120**
Total chlorophyll content (mg g^{-1} FW)	1.66 ± 0.023^{d}	1.67 ± 0.019^{d}	1.74 ± 0.017^{c}	1.81 ± 0.021^{b}	1.90 ± 0.025^{a}	1.87 ± 0.023^{ab}
Total carotenoids content (mg g^{-1} FW)	0.381 ± 0.016^{c}	0.383 ± 0.020^{c}	0.412 ± 0.018^{bc}	0.448 ± 0.017^{ab}	0.476 ± 0.022^{a}	0.455 ± 0.018^{ab}
CA activity (μmol CO_2 kg^{-1} (FW) s^{-1})	243.51 ± 2.41^{e}	239.51 ± 2.52^{e}	258.18 ± 2.54^{d}	279.71 ± 2.95^{c}	294.48 ± 2.59^{a}	286.07 ± 2.34^{b}
NR activity (n mol NO_2− g^{-1} FW h^{-1})	348.67 ± 3.24^{d}	350.67 ± 3.04^{d}	360.43 ± 3.52^{c}	378.75 ± 3.63^{b}	394.21 ± 3.19^{a}	387.48 ± 3.51^{b}
Leaf-nitrogen content (%)	2.64 ± 0.027^{e}	2.64 ± 0.027^{e}	2.73 ± 0.023^{d}	2.84 ± 0.029^{c}	3.05 ± 0.032^{a}	2.93 ± 0.028^{b}
Leaf-phosphorus content (%)	0.237 ± 0.004^{d}	0.235 ± 0.003^{d}	0.244 ± 0.003^{cd}	0.261 ± 0.003^{bc}	0.272 ± 0.004^{a}	0.249 ± 0.002^{ab}
Leaf-potassium content (%)	2.38 ± 0.029^{c}	2.37 ± 0.027^{c}	2.42 ± 0.031^{c}	2.59 ± 0.025^{a}	2.73 ± 0.035^{ab}	2.68 ± 0.027^{b}

*Means within a row followed by the same letter(s) are not significantly different ($p \leq 0.05$).

Note: Each value represents the mean of five replicates with ± S.E.

ISA: Irradiated sodium alginate; USA: Un-irradiated sodium alginate.

8.3.2.2 CARBONIC ANHYDRASE (CA) AND NITRATE REDUCTASE (NR) ACTIVITIES

Plant enzymes constitute a class of exceptional biomolecules with adroitly tailored ligand binding and signal perception mechanisms that control miscellaneous biosynthetic pathways, including phytohormone as well as bioactive compounds of secondary nature [12]. NR is a N assimilating enzyme which serves as a nutritional status indicator and stimulates the production of nitrite from nitrate by reversible reduction using NAD(P)H [19]. Likewise, CA is chiefly associated with augmenting the accessibility of CO_2 to RuBisCo and catalyzes HCO_3 formation. CA, being present in almost all photosynthesizing tissues, signifies its utmost significance in the life-sustaining process. ISA-induced upregulation in the activity of CA can be attributed to enhanced CO_2 diffusion through the stomatal apertures, which, in turn, is due to augmented stomatal conductance [40]. Significant increase in the enzymatic activities (NR and CA) of peppermint by graded ISA concentrations compared to the Control was observed; however, a maximum enhancement of 20.9% and 13.1% in the CA and NR activities, respectively, by 90 mg L^{-1} ISA over the Control was reported. The increase was progressive up to 90 mg L^{-1} ISA, ahead of which the values decreased (Table 8.2). ISA-stimulated upregulation in the nitrate assimilating enzyme, as observed here, can be endorsed to the enriched leaf N content. Increased N content augments the content of leaf nitrate to be incorporated in the biomolecules after its reduction [2, 3, 42]. Increased substrate concentration stimulates increased biosynthesis of functional enzymatic protein, possibly by synthesizing a mysterious nitrate-sensing molecule [19]. Moreover, upregulation of the vital gene-expression steps, notably transcription/translation due to ISA application, can lead to *de novo* enzyme biosynthesis, thereby increasing the fresh enzyme concentrations in the plant [40]. Oligosaccharides-induced enhancement in the activities of enzymes finds agreement with a number of studies.

8.3.2.3 LEAF-NITROGEN, PHOSPHORUS, AND POTASSIUM CONTENTS

Oligosaccharides can amplify permeability of biological membranes similar to conventional phytoregulators [3, 6]. The increased permeability facilitates nutrient absorption and assimilation, translocation of photosynthates, and their consumption, which eventually leads to improved biomass accumulation, as is evident in the current study from augmented dry weight of the plant [2]. Leaf-nitrogen, phosphorus, and potassium contents were maximally enhanced by the foliar application of 90 mg L^{-1} ISA as compared to the Control. An increase of 15.5% in leaf-nitrogen content, 14.8% in leaf-phosphorus content, and 14.7% leaf-potassium content over that of the Control was reported by this treatment (Table 8.2). A total of 90 mg L^{-1} ISA proves to be the optimum ligand concentration to enhance the nutrient absorption and assimilation maximally. Oligosaccharide-induced augmentation in growth probably enhances soil-nutrient uptake, ensuing their accretion in the peppermint leaves [5]. Oligomers of sodium alginate increase the activities of vital enzymes, as observed in the current study as well, which augment nutrient absorption and assimilation, thereby leading to improved photosynthesis and subsequent photosynthate-translocation and partitioning, improve the growth of different plant parts.

8.4 CONTENT AND YIELD OF ESSENTIAL OIL AND ACTIVE CONSTITUENTS (MENTHOL, MENTHONE, AND MENTHYL ACETATE)

Oligosaccharides have been reported to augment the photosynthetic parameters, notably content of pigments (total chlorophyll and carotenoids), chlorophyll fluorescence, stomatal conductance, net-photosynthetic rate and activities of rubisco and CA, which put in more photosynthates to be stored and diverted toward secondary metabolism [5, 6, 40]. Secondary metabolites, particularly EO, were markedly enhanced by leaf-applied ISA oligomers. Significant enhancement was observed in EO content as well as EO yield per plant. The treatment 90 mg L^{-1} ISA proved most effective and significantly enriched the content and yield of EO by 32.4% and 68.0%, respectively, over the Control. The effect of this treatment was followed by that of 120 mg L^{-1} ISA in case of both parameters. The values given by all the treatments were found to be significantly different from each other, and the lowest value for these two parameters was observed for the Control (Table 8.3). The increase in the content of secondary metabolites has also been endorsed to the increased activities of vital secondary metabolism enzymes like DXR and PAL, which consequently enrich the biosynthesis of phenolics and EOs . Increased expression of limonene synthase in the chloroplasts due to oligosaccharide application increases the content of EO in the chloroplast by speeding up the methyl-erythritol pathway [5, 29].

TABLE 8.3 Effect of Foliar Spray of ISA on Contents and Yields of Essential Oil, Menthol, Menthone and Menthyl Acetate of Peppermint After 150 dap (Days After Planting)*

Parameters	ISA Concentrations (mg L−1)					
	0	**USA**	**ISA-30**	**ISA-60**	**ISA-90**	**ISA-120**
Essential oil content (%)	0.68 ± 0.015^{e}	0.69 ± 0.021^{e}	0.73 ± 0.012^{d}	0.78 ± 0.017^{c}	0.90 ± 0.011^{a}	0.85 ± 0.017^{b}
Essential oil yield per plant (mL)	0.341 ± 0.015^{e}	0.352 ± 0.017^{e}	0.398 ± 0.012^{d}	0.443 ± 0.014^{c}	0.573 ± 0.013^{a}	0.517 ± 0.015^{b}
Menthol content (%)	34.92 ± 0.37^{b}	34.90 ± 2.62^{c}	35.09 ± 0.34^{ab}	35.55 ± 0.29^{ab}	36.09 ± 0.38^{a}	35.68 ± 0.28^{ab}
Menthol yield per plant (mL)	0.119 ± 0.010^{d}	0.121 ± 0.010^{d}	0.139 ± 0.009^{cd}	0.158 ± 0.011^{bc}	0.207 ± 0.013^{a}	0.185 ± 0.010^{ab}
Menthone content (%)	17.64 ± 0.301^{b}	17.84 ± 0.231^{b}	20.57 ± 0.255^{a}	20.15 ± 0.326^{a}	16.53 ± 0.225^{c}	17.35 ± 0.240^{bc}
Menthone yield per plant (mL)	0.060 ± 0.009^{b}	0.063 ± 0.011^{b}	0.078 ± 0.008^{ab}	0.089 ± 0.010^{ab}	0.095 ± 0.008^{a}	0.090 ± 0.009^{ab}
Menthyl-acetate content (%)	12.32 ± 0.222^{b}	12.29 ± 0.243^{b}	12.17 ± 0.208^{b}	12.97 ± 0.297^{b}	14.40 ± 0.277^{a}	14.92 ± 0.287^{a}
Menthyl-acetate yield per plant (mL)	0.042 ± 0.006^{c}	0.043 ± 0.006^{c}	0.046 ± 0.007^{c}	0.058 ± 0.008^{bc}	0.083 ± 0.008^{a}	0.077 ± 0.007^{ab}

*Means within a column followed by the same letter(s) are not significantly different ($p \leq 0.05$).

Note: Each value represents the mean of five replicates with ± S.E. Means within a row followed by the same letter are not significantly different from each other.

ISA: Irradiated sodium alginate; USA: Un-irradiated sodium alginate.

Moreover, oligosaccharide-induced upregulation of some monoterpenes metabolic enzymes may increase their contents [5]. Menthol content and yield were significantly enhanced by 90 mg L^{-1} ISA with an increment of 3.4% in comparison to DIW-sprayed Control. A maximum increase of 74% in menthol yield per plant was reported by 90 mg L^{-1}. A progressive increase up to 90 mg L^{-1} ISA was found, after which a decrease was reported in the values of both parameters. Foliar applied 90 mg L^{-1} ISA, however, significantly decreased the menthone content over the Control by 6.7%. The menthone yield per plant, being a computed value, was maximally enhanced by 58.3% due to 90 mg L^{-1} ISA over the Control because of the total increase in the EO yield per plant (Table 8.3). Higher concentrations of ISA (90 mg L^{-1} ISA and 120 mg L^{-1} ISA) significantly enhanced the menthyl acetate content compared to the Control, though both the treatments were at par. A maximum increase of 21.1% was noted in the menthyl acetate content by 120 mg L^{-1} ISA in comparison to the Control. The highest value in the menthone yield was observed by 90 mg L^{-1} ISA, which registered an increase of 97.6% over the Control (Table 8.3). Increased content of menthol may decrease the content of menthone or menthyl acetate. A similar observation has been reported in Japanese mint previously [5, 20, 21]. The current finding is in agreement with Ahmad et al. [5] in peppermint. Oligosaccharide-induced enhancement in the content of menthol, as has been reported in peppermint in the current study, might be attributed to the increased expression of menthol-reductase, and the possible conversion of one monoterpene into the other may result in the increase in the one at the cost of the other [5]. Upregulation of menthol reductase might be a reason for increased accretion of menthol [45]. Conclusively, sodium alginate oligosaccharides improve the performance of peppermint by improving growth, expression of vital enzymes, nutrient assimilation, photosynthesis, and EO accretion.

KEYWORDS

- **essential oil**
- **polysaccharide**
- **sodium alginate**
- **peppermint**
- **deionized water**

REFERENCES

1. Abad, L.V., Aurigue, F.B., Relleve, L.S., Montefalcon, D.R.V., Lopez, G.E.P. Characterization of low molecular weight fragments from gamma irradiated κ carrageenan used as plant growth promoter. *Radiation Physics and Chemistry*, 2016, *118*, 75–80. https://doi.org/10.1016/j.radphyschem.2015. 03.001
2. Aftab, T., Khan, M.M.A., Naeem, M., Idrees, M., Varshney, L. Enhancing the growth, photosynthetic capacity and artemisinin concentration in *Artemisia annua* L. by irradiated sodium alginate—A plant growth promoter. *Radiation Physics and Chemistry*, 2011, *80*, 833–836.
3. Aftab, T., Naeem, M., Direst, M., Khan, M.M.A., Moinuddin, Varshney, L.S. Cumulative role of irradiated sodium alginate and nitrogen fertilizer on growth, biochemical processes and artemisinin production in *Artemisia annua*. *Industrial Crops and Products*, 2013, *50*, 874–881.
4. Aftab, T., Khan, M.M.A., Naeem, M., Idrees, M., Siddiqi, T.O., Varshney, L. Effect of irradiated sodium alginate and phosphorus on biomass and artemisinin production in Artemisia annua. *Carbohydrate Polymers*, 2014, *110*, 396–404.
5. Ahmad, B., Khan, M.M.A., Jaleel, H., Sadiq, Y., Shabbir, A., Uddin, M. Exogenously sourced γ-irradiated chitosan-mediated regulation of growth, physiology, quality attributes and yield in mentha piperita L. *Turkish Journal of Biology*, 2017, *41*, 388–401. https://doi.org/10.3906/biy-1608-64
6. Ahmad, B., Dar, T.A., Khan, M.M.A., Ahmad, A., Rinklebe, J., Chen, Y., Ahmad, P. Oligochitosan fortifies antioxidative and photosynthetic metabolism and enhances secondary metabolite accumulation in arsenic-stressed peppermint. *Frontiers in Plant Science*, 2022, *13*, 987746. https://doi.org/10.3389/fpls.2022.987746
7. Ahmad, B., Jahan, A., Sadiq, Y., Shabbir, A., Jaleel, H., Khan, M.M.A. Radiation-mediated molecular weight reduction and structural modification in carrageenan potentiates improved photosynthesis and secondary metabolism in peppermint (*Mentha piperita* L.). *International Journal of Biological Macromolecules*, 2019a, 124, 1069–1079. https://doi.org/10.1016/j.ijbiomac.2018.12.022
8. Ahmad, B., Jaleel, H., Shabbir, A., Khan, M.M.A., Sadiq, Y. Concomitant application of depolymerized chitosan and GA3 modulates photosynthesis, essential oil and menthol production in peppermint (*Mentha piperita* L.). *Scientia Horticulturae*, 2019b, 246, 371–379. https://doi.org/10.1016/j.scienta.2018.10.031
9. Ahmadi, A., Moghadamtousi, S.Z., Abubakar, S., Zandi, K. Antiviral potential of algae polysaccharides isolated from marine sources: A review. *BioMed Research International*, 2015, *2015*, 825203.
10. Al-Assaf, S. Phillips, G.O. Williams, P.A. Du Plessis, T.A. Application of ionizing radiations to produce new polysaccharides and proteins with enhanced functionality. *Nuclear Instruments and Methods in Physics Research Section B: Beam Interactions with Materials and Atoms*, 2007, *265*, 41–42.
11. Castro, J. Vera, J. González, A. Moenne, A. Oligo-carrageenans stimulate growth by enhancing photosynthesis, basal metabolism, and cell cycle in tobacco plants (var. burley). *Journal of Plant Growth Regulation*, 2012, *31*, 173–185

12. Chmielewski, A.G., Migdal, W., Swietoslawski, J., Jakubaszek, U., Tarnowski, T. Chemical radiation degradation of natural oligoamino-polysaccharides for agricultural application. *Radiation Physics and Chemistry*, 2007, *76*, 1840–1842.
13. Dar, T.A., Uddin, M., Khan, M.M.A., Ali, A., Varshney, L. Modulation of alkaloid content, growth and productivity of Trigonellafoenum-Graecum L. using irradiated sodium alginate in combination with soil applied phosphorus. *Journal of Applied Research on Medicinal and Aromatic Plants*, 2016, *3*(4), 200–210. https://doi.org/10.1016/j.jarmap.2016.05.003
14. de Jesus Raposo, M.F., deMorais, A.M.B., deMorais, R.M.S.C. Marine polysaccharides from algae with potential biomedical applications. *Marine Drugs*, 2015, *13*, 2967–3028, https://doi.org/10.3390/md13052967.
15. Dwivedi, R.S., Randhawa, N.S. Evaluation of rapid test for the hidden hunger of zinc in plants. *Plant and Soil*, 1974, *40*, 445–451.
16. El-Rehim, A.H.A. Characterization and possible agricultural application of polyacrylamide/sodium alginate crosslinked hydrogels prepared by ionizing radiation. *Journal of Applied Polymer Science*, 2006, *101*, 3572–3580.
17. Guenther, E. The essential oils: History, origin in plants, production. *Analysis*, 1972, *1*, 147–151.
18. Hald, P.M. The flame photometer for the measurement of sodium and potassium in biological materials. *Journal of Biological Chemistry*, 1947, *167*, 499–510.
19. Hewitt, E.J, Afridi, M.M.R.K. Adaptive synthesis of nitrate reductase in higher plants. *Nature*, 1959, *183*, 57–59.
20. Jaleel, H., Khan, M.M.A., Ahmad, B., Shabbir, A., Sadiq, Y., Uddin, M., Varshney, L. Elucidating the dynamics of physiology and essential oil production in lemongrass using alginate oligomers under field conditions. *Journal of Functional and Environmental Botany*, 2017a, *7*(1), 26–35. https://doi.org/10.5958/2231-1750.2017.00004.X
21. Jaleel, H., Khan, M.M.A., Ahmad, B., Shabbir, A., Sadiq, Y., Uddin, M., Varshney, L. Essential oil and citral production in field-grown lemongrass in response to gamma-irradiated chitosan. *Journal of Herbs Spices & Medicinal Plants*, 2017b, 10, 1–5.
22. Jaworski, E.G. Nitrate reductase assay in intact plant tissues. *Biochemical and Biophysical Research Communications*, 1971, *43*, 1247–1279.
23. Lichtenthaler, H.K., Buschmann, C. Chlorophylls and carotenoids: Measurement and characterization by UV-VIS spectroscopy. *Current Protocols in Food Analytical Chemistry*, 2001, *1*, F4. 3.1–F4. 3.8. https://doi.org/10.1002/0471142913.faf0403s01
24. Lindner, R.C. Rapid analytical methods for some of the more common inorganic constituents of the plant tissues. *Plant Physiology*, 1944, *19*, 76–89.
25. Luan, L.Q., Nagasawa, N., Tamada, M., Nakanishi, T.M. Enhancement of plant growth activity of irradiated chitosan bymolecular weight fractionation. *Radioisotopes*, 2006, *55*, 21–27.
26. Mahendran, G., Rahman, L.U. Ethnomedicinal, phytochemical and pharmacological updates on Peppermint (Mentha× piperita L.)—A review. *Phytotherapy Research*, 2020, *34*, 2088–2139. https://doi.org/10.1002/ptr.6664
27. Márquez-García, B., Fernández-Recamales, M., Córdoba, F. Effects of cadmium on phenolic composition and antioxidant activities of Erica andevalensis. *Journal of Botany*, 2012, *2012*, 936950. https://doi.org/10.1155/2012/936950

28. Mimica-Dukic, N., and Bozin, B. Mentha L. species (Lamiaceae) as promising sources of bioactive secondary metabolites. *Current Pharmaceutical Design*, 2008, *14*, 3141–3150. https://doi.org/10.2174/138161208786404245
29. Mollah, M.Z.I., Khan, M.A., Khan, R.A. Effect of gamma irradiated sodium alginate on red amaranth (*Amaranthus cruentus L.*) as growth promoter. *Radiation Physics and Chemistry*, 2009, *78*, 61–64.
30. Naeem, M., Idrees, M., Aftab, T., Khan, M.M.A., Moinuddin, Varshney, L. Irradiated sodium alginate improves plant growth, physiological activities, and active constituents in *Mentha arvensis* L. *Journal of Applied Pharmaceutical Science*, 2012, *2*, 28–35.
31. Naeem, M., Nabi, A., Aftab, T., Khan, M.M.A. Oligomers of carrageenan regulate functional activities and artemisinin production in *Artemisia annua* L. exposed to arsenic stress. *Protoplasma*, 2020, *257*(3), 871–887. https://doi.org/10.1007/s00709-019-01475-y
32. Naikoo, M.I., Dar, M.I., Raghib, F., Jaleel, H., Ahmad, B., Raina, A., et al. Role and regulation of plants phenolics in abiotic stress tolerance: An overview. *Plant Signaling Molecules*, 2019, 157–168. https://doi.org/10.1016/B978-0-12-816451-8.00009-5
33. Nayak, P., Kumar, T., Gupta, A.K. and Joshi, N.U. Peppermint a medicinal herb and treasure of health: A review. *Journal of Pharmacognosy and Phytochemistry*, 2020, *9*(3), 1519–1528.
34. Novozamsky, I., Houba, V.J.G., van Eck, R., van Vark, W. A novel digestion technique for multi-element plant analysis. *Communications in Soil Science and Plant Analysis*, 1983, *14*, 239–248
35. Pharmacopoeia, E. *Determination of Essential Oils in Herbal Drugs*. Council of Europe Strasbourg Cedex: France. 2008.
36. Pytlakowska, K., Kita, A., Janoska, P., Połowniak, M., Kozik, V. Multi-element analysis of mineral and trace elements in medicinal herbs and their infusions. *Food Chemistry*, 2012, *135*, 494–501. https://doi.org/10.1016/j.foodchem.2012.05.002
37. Remminghorst, U., Rehm, B.H. Bacterial alginates: From biosynthesis to applications. *Biotechnology Letters*, 2006, *28*, 1701–1712.
38. Ribeiro, M.A., Oikawa, H., Mori, M.N., Duarte, C.L. Degradation mechanism of polysaccharides on irradiated sugarcane bagasse. *Radiation Physics and Chemistry*, 2013, *84*, 115–118.
39. Sachan, N.K. Jha, S. Pushkar, A. Bhattcharya, A. Sodium alginate: The wonder polymer for controlled drug delivery. *Journal of Pharmacy Research*, 2009, *2*, 1191–1199.
40. Sadiq, Y., Khan, M.M.A., Shabbir, A., Ahmad, B., Jaleel, H., Uddin, M., Varshney, L. Structural re-arrangement of depolymerized sodium alginate enriches peltate glandular trichomes and essential oil production of spearmint. *International Journal of Biological Macromolecules*, 2017, *105*(1), 1043–1050. https://doi.org/10.1016/j.ijbiomac.2017.07.134
41. Sen, M., Yolacan, B., Guven, O, Radiation-induced degradation of galactomannan polysaccharides, *Nuclear Instruments and Methods B*, 2007, *265*, 429–433.
42. Shabbir, A., Khan, M.M.A., Sadiq, Y., Jaleel, H., Ahmad, B., Uddin, M. Regulation of functional activities and essential oil production in *Vetiveria zizanioides* L. Nash

after γ-irradiated sodium alginate elicitation. *Turkish Journal of Biology*, 2017, *41*, 661–672.
43. Stone, B.A. Svensson, B. Collins, M.E. Rastall, R.A. Polysaccharide degradation. *Glycoscience: Chemistry and Chemical Biology*. 2008; pp. 2325–2375.
44. Szymczycha-Madeja, A., Welna, M., Zyrnicki, W. Multi-element analysis, bioavailability and fractionation of herbal tea products. *Journal of the Brazilian Chemical Society*, 2013, *24*, 777–787. https://doi.org/10.5935/0103-5053.20130102
45. Turner, G.W., Croteau, R. Organization of monoterpene biosynthesis in *Mentha*. Immunocytochemical localizations of geranyl diphosphate synthase, limonene-6-hydroxylase, isopiperitenol dehydrogenase, and pulegone reductase. *Plant Physiology*, 2004, *136*, 4215–4227.
46. Uribe, E., Marín, D., Vega-Gálvez, A., Quispe-Fuentes, I., Rodríguez, A. Assessment of vacuum-dried peppermint (*Mentha piperita* L.) as a source of natural antioxidants. *Food Chemistry*, 2016, *190*, 559–565. https://doi.org/10.1016/j.foodchem.2015.05.108

CHAPTER 9

Traditional and Pharmacological Significance of Peppermint and Its Secondary Metabolites

MUKUL MACHHINDRA BARWANT,[1*] BALWANT SINGH,[2] SAGAR BAWAKE,[3] VIDYA MHASKE,[3] MEHTAB YASMEEN,[4] NITIN TANAJI GORE,[1] MAHENDRA LAXMAN AHIRE,[1] and AJAR NATH YADAV[5]

[1]*Department of Botany, Yashavantrao Chavan Institute of Science, Satara, Maharashtra, India*

[2]*Department of Botany, K. S. Saket PG College, Dr Ram Manohar Lohia Avadh University, Ayodhya, Uttar Pradesh, India*

[3]*Department of Botany, Ahmednagar College, Ahmednagar, Maharashtra, India*

[4]*Telangana Mahila Viswavidyalayam, Hyderabad, Telangana, India*

[5]*Department of Genetics, Plant Breeding, and Biotechnology, Dr Khem Singh Gill Akal College of Agriculture, Eternal University, Sirmour, Himachal Pradesh, India*

*Corresponding author

ABSTRACT

The Lamiaceae, commonly known as the mint family, is among the largest botanical families, comprising approximately 6900 to 7200 recognized species distributed across 236 genera. The most significant naturally occurring antioxidants, phenolic chemicals like flavonoids, are nearly

Peppermint: Novel Strategies for Performance Improvement. Bilal Amhad, Tariq Ahmad Dar, and Khalid Rehman Hakeem (Eds.)

DOI: 10.1201/9781779641557-9

exclusively found in plants. As a hybrid of two different flowering plant species from the dark Lamiaceae, mint genus, and order of the Tubliforal family Lamiaceae, *Mentha piperita* L. is a unique herb. The family has long been prized for its culinary, aromatic, medicinal, and environmental qualities. Mint family plants are useful in ancient medicine around the world. Popular plant family herb *M. piperita* is used to make food, cosmetics, personal care items, and other pharmaceutical products. It is grown for its flavoring and aroma qualities. Mint species are commonly employed in both traditional and modern medicine because they include two groups of secondary biomolecules: many structural types of phenolic compounds and monoterpenoids found in essential oils. Particular lipophilic flavonoids are present in mints, which is how they are recognized. There is a broad spectrum of pharmacological activity linked to the phenolic chemicals found in mints, such as anti-inflammatory, cholagogue, cytoprotective, hepatoprotective, antioxidant, and antidiabetogenic effects. However, some mint species can have negative effects on human health in addition to having restorative powers. The present chapter deals with the traditional and pharmacological significance of peppermint and its secondary metabolites.

9.1 INTRODUCTION

Herbs classified as medicinal can be both natural products and have the potential to prevent disease. Owing to their therapeutic properties, these plants' phytochemicals are able to useful curing diseases [40, 56]. Mint is well-known for its flavor and scent. In India, peppermint is vernacularly known as pudina and globally famous for its cooling and calming effects [72]. Peppermint is a very ancient herb and mentioned in Ayurveda as medicine to maintain "vata dosha," "pitta dosha," and "kapha dosha." In actuality, *Mentha spicata* and *Mentha aquatica* are crossed-hybrids (GM-plants) that produce *Mentha piperita* [44, 84]. For almost a millennium, this hybrid species has been raised in temperate, tropical, and subtropical climates [28, 137]. Peppermint comes in two varieties: white and black as mentioned in the literature [28]. The peppermint plant is now 60–90 cm taller, with dark green, uniquely scented leaves. It is a perennial herb with serrated, transverse and elliptic leaves. Its oil contains menthone and carboxyl esters, and its menthol level is exceptionally high. It appears that there must be pharmacological and popular traditional meanings for this [90].

Peppermint is a commonly grown, supplied, and sought-after plant worldwide due to its use as a raw ingredient in the pharmaceutical, culinary, and cosmetics industries [3, 44, 132]. The basic product of peppermint is its oil, which is produced from the leaf extraction and is used in different industries like food, flavor, pharmaceutical, and perfumery [89, 132, 134]. The prime constituent of peppermint oil is "menthol" and is widely used in the manufacturing of soap, lozenges, shampoo, toothpaste, cold balms, pain balms, massage oil, etc [71, 132]. Additionally, several stomach conditions, including indigestion, acidity, gas problems, etc., are treated with mint oil [3, 132]. Menthol is used to make different ayurvedic medicines, like very famous one "Dabur Pudin Hara" and used as a chief ingredient of cough drops, cough candy (e.g., Strepsil, Vicks), and ointments (e.g., Vicks Vaporub) [3, 132]. Peppermint extracts have several distinct inhibitory effects like antimicrobial, antioxidant, antinausea, antispasmodic, etc., especially in gastrointestinal motility [15, 53, 59, 66, 108].

The chemical compounds of plants produced from the primary metabolic pathway are known to be secondary metabolites. Over 2,000,000 secondary metabolites in plants are known with their different functions in the plant kingdom [30]. In the glandular trichomes of plants, particularly leaves, peppermint oil is produced and stored as complex combinations of secondary metabolites. The distinctive scents are mostly caused by secondary metabolites made of volatile mono- and sesquiterpenes. Naturally, environmental elements such as temperature, humidity, light intensity, soil quality, and water availability affect the number and quality of secondary metabolites [37, 77, 80, 93]. Because of these secondary metabolites' aromatic, medicinal, and culinary qualities, they are utilized as food additives and medications [8, 155].

Allelopathy refers to any mechanism involving secondary metabolites generated by bacteria, fungi, plants, algae, or fungi that affects the growth and development of biological systems. Allelochemicals make up all secondary metabolites [12]. Herbal medications use low molecular weight and active natural secondary metabolites from plants to distinguish between human diseases and syndromes. Based on chemical structure, secondary metabolites are diversified into two basic categories, that is, nitrogenous and non-nitrogenous secondary metabolites [93]. Primary metabolites play a role in forming the molecular skeleton, whereas secondary metabolites play an important role in defense and self-defense functions [138].

Several pharmacological properties of peppermint (Figure 9.1) have been reported in many scientific literatures, including antiallergenic, antimicrobial activities, chemopreventive potential, hepatic and renal actions, nervous system actions, antioxidant activity, antitumor, antispasmodic, gastrointestinal effects, anti-inflammatory, anti-irritant, pulmonary effects, and fatigue of cardiac patients [42, 66, 88, 102]. Other than pharmacological, medicinally also peppermint is very important and useful for human health and cures several diseases (Figure 9.2).

FIGURE 9.1 Pharmacological properties of peppermint.

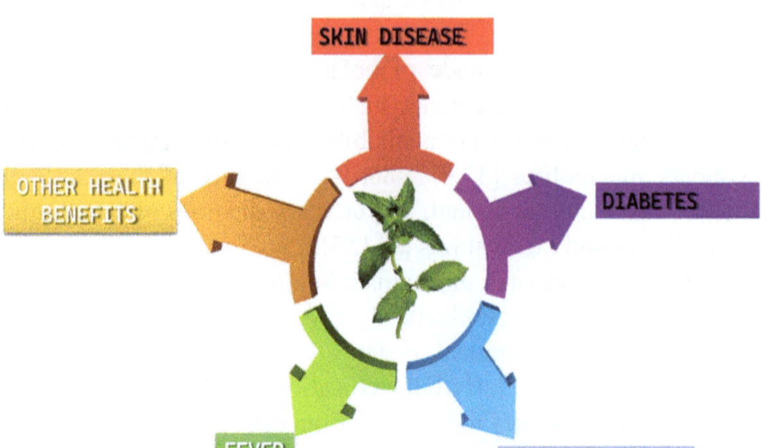

FIGURE 9.2 Medicinal properties of peppermint.

9.2 BIOACTIVE COMPOUNDS OF PEPPERMINT

Peppermint oil chemical components are incredibly complex and varied. Relative concentrations are impacted by several elements, such as climate, cultivar, geographic location, various environmental factors (soil properties, moisture content, and temperature), phenological status (plant stage at which it is harvested), the plant's parts (flowers, stems, leaves, or the entire plant), the kind of material (fresh or dried), and even the techniques employed in the process of extracting essential oils [86]. Although there are other chemicals as well, pharmaceutically significant "bioactive compounds," which are classified as secondary metabolites into two categories: essential oils and phenolic compounds.

9.2.1 ESSENTIAL OILS

Peppermint essential oils act as antispasmodic, carminative, antibacterial, and antiviral agents. They are the primary bioactive components of peppermint. In addition, several kinds of mint oil are thought to contain naturally occurring antioxidants. By using a hydrodistillation method, dried mint leaves or mint leaves straight from the garden can be used to extract the essential oil. The primary hydrocarbon chemicals found in peppermint oil are phenylpropanoids, oxygen derivatives, and monoterpene and sesquiterpene hydrocarbons [2]. One of peppermint main ingredients, menthol, is frequently used to treat nasal congestion [52], congestion, and headache issues [31]. Peppermint oil has several external uses, including treating rheumatism, toothaches, congestive headache, neuralgia, and other conditions due to its inherent antibacterial and anesthetic properties. Menthol is utilized as a calming balm and is often referred to as "Peppermint Camphor." Peppermint extracts are typically added to cigarettes as an addition to mask the sour and bitter taste of tobacco. This relieves sore throats without tasting harsh. Peppermint leaves have been shown to contain 1.2% oils, with at least 300 different components discovered. With 52% of monoterpenes and 9% of sesquiterpenes, the terpenic class is one of the most abundant; other groups with smaller representations include 9% of aromatic hydrocarbons [49].

The composition of peppermint oil varies widely, and the chemorace or chemotype has a significant impact on the pharmacological efficacy [48].

Additionally, it was discovered that the primary ingredient in peppermint essential oils made from material gathered from plants in Saudi Arabia, China, Iran, Taiwan, and Brazil was menthol [99, 120, 141, 142, 148]. This species has another chemotype that is distinguished by a high concentration of carvone as the primary ingredient, followed by limonene [126].

9.2.2 PHENOLIC COMPOUNDS

Within species, the composition of phenolic components is comparatively consistent. Flavonoids are the most significant phenolic chemicals found in *Mentha* species. A particular type of lipophilic flavonoid is present in mints. Flavonoid glycosides and polyphenols are examples of phenolic components [13, 150]. There are several distinct components in the various species of mint, including glycosides, acylated flavonoids, cinnamic acids, and/or aglycon [50]. Additionally, it was discovered that the leaf extract included lignans and stilbenes [89]. Numerous other compounds can also be found in peppermint [110, 144]. Leaf tissue was found to contain flavones such as hesperidin, naringenin, and eriodictyol [50]. [64]. Generally speaking, leaves' glandular trichomes contain the essential oils [115]. Triantaphyllou et al. [140] discovered that the extracts of mint contained esters of hydroxylated glycosidic flavonoids, phenolic carboxylic acids, and other flavonoid derivatives. Salvianolic acid is among the approximately seven salvianolic acids found in mint plants [73]. Extracts of dried water mint, spearmint, peppermint, and bergamot mint have been shown in earlier research to contain lipophilic methylated flavonoids. Currently, 20 flavonoids are identified. It has been observed that 5,6-dihydroxy-7,8,3′,4′-tetramethoxyflavone is the primary bioflavonoid found in peppermint [150].

9.2.3 OTHER COMPONENTS

According to studies, mentha also contains additional substances. Reduced concentrations of several additional substances are also present in peppermint. Linoleic, linolenic, and palmitic acids are the main compounds present in peppermint leaves [112]. Various species of mint contain different kinds of colors. There have been reports of pigments such as chlorophyll a and b [43, 117]. The peppermint tea contains isomers of lutein and β-carotene, but also lutein in combination. Numerous vitamins, including C and E, were found in mint [121]. Compounds such as quinines, anthraquinones, saponins, and alkaloids were also found in numerous investigations [109].

9.3 NUTRITIVE SCENARIO OF PEPPERMINT

A perennial member of the Lamiaceae family, mint (*M. spicata* L.) grows in semiarid conditions all over the world [14]. Mint is a lucrative crop that produces essential oil in addition to fresh or dried leaves. Herbal teas can be made with both dried and fresh leaves. Essential oils are extracted from the entire plant and used in food products, cosmetics, nutraceuticals, and pharmaceuticals [14–79]. In recent years, Mexico has greatly expanded the production of fragrant herbs for fresh export; yet, in order to be exported, the herbs must meet strict quality standards that are controlled during manufacture. Applying fertilizer and providing adequate watering are the two most crucial agronomic factors affecting the productivity of mint crops [157]. Mineral nutrients are necessary for mentha plants to develop and produce essential oils [139].

Peppermint oil is made from the leaves of two perennial plants in the Labiatae family: *M. piperita* L. and *Mentha arvensis* var. piperascens. India is the largest producer and exporter of mint oil worldwide. Mint oil and its constituents find application in the food, pharmaceutical, flavoring, and fragrance industries. Lozenges, toothpastes, cold and pain balms, Dabur Pudin Hara, and other items are made with menthol, which is the primary ingredient in these products. The leaves of the *M. arvensis* plant serve as the primary material [3]. The oil is used to treat gastrointestinal conditions like gas, acidity, indigestion, and more. Key component of Ayurvedic products like Dabur's "Pudin Hara." Menthol, which is used in cough drops and ointments like Vicks Vaporub, is naturally sourced from the oil. The food industry's use of *Mentha* spp. will offer a cost-effective and biocompatible method of managing diabetes and obesity [33]. Hyperglycemia, or elevated blood glucose levels brought on by insufficient or decreased insulin, is a hallmark of diabetes, a metabolic disease [11]. Mentha's nutritional qualities and capacity to replace sugar account for its extensive use and economic standing [116].

The risk of diabetes and cardiovascular disease can be reduced by using mentha phytoconstituents as food additives and preservatives. Because of high-fat diets and insufficient exercise, diabetes and cardiovascular disease are becoming more widespread worldwide [63]. The nutritional value of food is diminished by high triglyceride, artificial taste, and preservative levels, which are also associated with diabetes, obesity, and other chronic illnesses [45]. A total of 30–80% of people are at risk for diabetes and obesity as a result of their diet and activity habits [29]. Insulin tablets and eating sugar-free foods are two of the many strategies used to control DM [68–154]. Strategies have an adverse effect on patients' general quality of life, food satisfaction, and nutrition status [51–60]. In the past 20 years, natural methods of managing metabolic issues, such as nutrition, have been tried. Food processing uses a number of chemical treatments, but research has shown that many have unfavorable side effects, which has led to the use of active natural compounds [95].

Pure or contaminated plant-derived extracts offer countless options as nutritious and biocompatible food products [41–105]. Epidemiology researchers are increasingly recommending a wide variety of aromatic and therapeutic plants due to their nutritional and preservation qualities [103–123]. Humans can receive plant-based nourishment by consuming aqueous extracts of medicinal plants in their diets [19–67]. Usually, a physical method that does not alter the composition of the aqueous phase is used to make aqueous extracts [114]. Nevertheless, before using these extracts widely, a comprehensive examination of their cytotoxicity, anti-oxidant, antidiabetic activity, and lipid oxidation potential is required in order to confirm their safety and effectiveness through a proof-of-concept study for possible health claims [45–58].

Many ailments, such as pain management, gastrointestinal issues, obesity, anticarcinogenic properties, antifungal and antibacterial activities, antiallergic effects, vomiting, and nausea, have been treated for many years with mentha, a popular medical and economic plant [38–54]. To help reduce diabetes and obesity brought on by artificial food additives and preservatives, the primary characteristics of extracts from the genus *Mentha* that ought to be regarded as preservatives and food additives were mentioned. It was discovered that the tastes and phytochemicals present in *Mentha* species can be added to typical diet items to improve food quality in a sustainable and cost-effective manner. Additionally, studies on a range of aromatic and medicinal plants should be carried out, with a focus on their potential as dietary supplements and food additives, as well as their potential to treat fatal illnesses like diabetes and obesity through regular dietary products. When administered topically, peppermint extract raised people's pain thresholds [92].

Significant differences were observed in the peppermint group's self-evaluated performance, felt physical strain, and perceived temporal stress. In an animal experiment, injecting several peppermint components intraperitoneally into mice significantly increased their ambulatory activity. Consequently, the author postulated that the active components of peppermint stimulate the central nervous system [151]. Furthermore, ethanol and peppermint oil were found to have potent analgesic properties [65]. Plants can be utilized in many different ways to generate food for people and other animals, as well as raw materials for industry, but it is crucial to consider the nutritional value of plant resources. Equally vital raw materials are provided by plants to our industry. In addition to being widely used as a source of energy, plants are today the primary source of medicine, according to historical and scientific data. A popular fragrant and therapeutic herb, peppermint is employed in the pharmaceutical, cosmetic, and culinary industries worldwide. Of its components, menthol is the most prevalent and advantageous [44].

A number of physiological, medical, or dietary problems can cause hair loss, an unpleasant condition. Male pattern baldness, or androgenetic alopecia, for example, is becoming recognized as a medical condition requiring specialist care from generalist practitioners due to its considerable psychological and physical effects [140]. In order to treat hypertension, Upjohn Pharmaceuticals initially created minoxidil [131].

9.4 PHARMACOLOGICAL SIGNIFICANCE

The popular herb peppermint is prized for its invigorating flavor and aroma. It has been used for possible health benefits for centuries, and studies on its pharmacological properties have been conducted. The primary active ingredient in peppermint is menthol. The peppermint's traditional remedies for indigestion, gas, bloating, and other gastrointestinal discomforts. The digestive tract's smooth muscles are relaxed by menthol, potentially reducing spasms and fostering more efficient digestion. Irritable bowel syndrome may benefit from peppermint oil therapy, according to various research conducted by a lot of researchers. While in studying the pharmacological potential of *Mentha*, researchers have found tremendous use of

Mentha for mankind. Whenever we thought about the use of *Mentha* then we would get lots of things from its potential as the antibacterial as well as antifungal properties, its antioxidant nature, its antiviral potential, as well as its anti-inflammatory activity, and many more [9, 125].

9.4.1 ANTIOXIDANT ACTIVITY

In the cell, ROS is constantly created from oxygen, primarily in the respiratory chain of the mitochondria, which is indicative of a free radical assault on a cell and creates oxidative stress on the cell. Basically, ROS is a very toxic and reactive substance. Mentha is a plant which known for its aroma and diversity of various classes of biochemicals. A well-known family of compounds for their antioxidant properties is phenolic compounds.

This class of biochemicals is a secondary metabolite are of those who is under extensive research for its potential use to health benefit for mankind. Some of the important phenolic compounds is: rosmarinic acids, which are abundant in mint leaves, caffeic acid, and luteolin, which are flavone-type phenolic compounds [34]. Apigenin is also a flavone which more abundant, quercetin, as well as salvianolic acid, which is not common in all mint species, but it is known for its antioxidant nature and found in mint. These antioxidants aid in the body's defense against dangerous free radicals that can cause oxidative stress and injury to cells and tissues. The main antioxidants may be phenolic acids such as caffeic and rosmarinic acids, luteolin (flavones), and eriocitrin (flavanones).

Ascorbic acid and carotenoids, two vitamin antioxidants, represent minor enhancements in the overall capacity of antioxidants. Acyclic unsaturated oxygenated monoterpenes, like linalool, could possess pro-oxidant properties, have scavenging potential in essential oils [121]. L-menthone and pulegone are the main essential volatile compounds. And apigenin is the main flavonoid in methanolic extracts of mentha, which is well known for its antioxidant activity [133]. The mint leaves' essential oils were hydro-distilled to ascertain their capability for scavenging free radicals *in vitro* using the tests for the breakdown of deoxyribose and the DPPH assay. It is indicating robust radical scavenging capabilities [61]. In the DPPH and ABTS tests, aerial parts of mint species *spictata's* ethanolic extract have significant power [104]. Methanolic extracts of mint species were examined due to their ability to protect PC12 cells against hydrogen peroxide-induced toxicity [85].

9.4.2 ANTIVIRAL ACTIVITY

It was found how well peppermint oil worked as a virucidal agent against the herpes simplex virus. Peppermint oil had a potent virucidal effect in viral suspension assays against herpes simplex virus-1 and herpes simplex virus-2, forming plaques HSV-1 by 82% and HSV-2 was 92%, at noncytotoxic oil concentrations. Higher peppermint oil concentrations decreased both herpesvirus viral titers by more than 90% [129]. At noncytotoxic doses, water extracts from members of the Lamiaceae family can significantly and quickly decrease the infectiousness of HIV particles. The most likely mechanism of action appears to be the outcome of an extractive rise in the density of the virion before it meets the surface [62].

By evaluating the potential inhibitory effects of peppermint extracts and menthol, the antiviral activity on pretreatment cells, during viral adsorption, and after the viral adsorption phase could be determined using the plaque reduction test, a pure peppermint component, against both strains of HSV [156]. By the *in silico* screening of mint leaves showed that compounds like rutin, hesperidin, and isorhamnetin are present in mint leaves[130]. The *Mentha* possesses significant activity against the RSV, that is respiratory syncytial virus, which causes respiratory tract infection. Ethanolic extract of mint leaves has antiviral efficaciousness against RSV, with value of IC_{50} is 10.41 μg/mL [82]. The recent research finding suggested that the various monoterpenes like menthofuran, 1,8-cineol, and pulegone from species of *Mentha* which is having potential action against the SARS-CoV-2 viral strains [39].

9.4.3 ANTIBACTERIAL ACTIVITY

The oil produced from the peppermint is a secondary metabolite. Due to this, peppermint shows antibacterial, antifungal activity, as well as shows biological activity against various bacterial strains. The antibacterial qualities can affect both gram-positive and gram-negative microorganisms of mint essential oil. Mint essential oil has uses in pharmacological, agricultural, food, cosmetic, hygienic, and perfusion fields. One of the most often utilized flavors around the globe is mint oil, which is a key ingredient in many cuisines and confections. Menthol, menthone, and neomenthol are the ingredients found in peppermint oil. There have been reports of a variety of biological activities, including antibacterial ones, in peppermint oil and its components. *Staphylococcus aureus* and *Staphylococcus pyogenes*, two Gram +ve bacteria strains, are susceptible to mint oil, according to the author most sensitive strains were discovered to *S. aureus*, *S. pyogenes,* as well as *Klebsiella pneumoniae* having discovered to be more sensitive to essential oil than *Escherichia coli.*

During the antibacterial testing, several solvents were more efficient to dissolve biological components, that is, petroleum ether extracts, ethyl acetate extract, and chloroform extract were discovered to be greater efficient than watery and alcoholic extracts. According to results, ethyl acetate extract and pet ether extract had greater efficiency against *S. aureus* and *K. pneumoniae* than against *S. pyogenes* and *E. coli* among the examined pathogens [134]. The primary constituents of *Mentha pulegium* are essential oils (71.1%) were discovered to be piperitone and piperitenone, which also show considerable antibacterial action. It has been demonstrated that the antibacterial qualities of the essential oils of mint.

Historic applications of the two mentha species for treating microbe-related ailments elucidated by the high amounts of carvone and pulegone [32]. At varying concentrations (1:1, 1:5, 1:10, and 1:20), extracts of mentha species made with water as a solvent are effective against a variety of bacteria, including effective against *Salmonella choleraesuis*, *E. coli*, *Pseudomonas aeruginosa*, *K. pneumoniae*, *Bacillus fastidiosus*, *S. aureus*, *Proteus mirabilis*, *Proteus vulgaris*, and *Serratia odorifera* [10]. *Bacillus cereus, Salmonella enterica* subsp. enterica serovar Typhimurium, and *Staphylococcus epidermidis* were all inhibited in growth by mint essential oil [57]. In order to pinpoint the precise mechanism of mint leaves' antimicrobial activity, it will be crucial in the future from a scientific extract, potent ingredient from mint and carry out several bioassays and clinical investigations. Studies are being done on the therapeutic potential of the plant's seeds, flowers, and other sections in addition to its leaves and bark [75].

9.4.4 ANTIFUNGAL ACTIVITY

The essential oil's antifungal qualities are demonstrated against 25 common fungi, such as *Candida* spp. Also investigated the EO's antifungal effects on 35 clinical yeast isolates determined by the PCR–RFLP method [98]. The ability of certain organic extracts (rich in bioactive chemicals) to stop the development of numerous strains linked to human illnesses, including *Microsporum audouinii*, *Aspergillus niger*, *Cryptococcus neoformans*, and *Candida albicans* exhibited significant antifungal activity [70]. Several researches used various plant parts and techniques, including disc diffusion, microdilution broth susceptibility assay, agar well diffusion and spots to examine the efficacy of mint extracts on fungi [6, 70, 83, 107]. Five quantities of mint oil were used in the agar dilution procedure to assess the antifungal assay's results. Under laboratory testing, it was discovered that the oil was efficient against certain fungal infections. With an increase in concentration, the oil's antifungal properties grew. The minimum oil concentrations required to effectively combat fungi were also variable [99].

Strong effectiveness was shown in spearmint against molds that cause food to spoil [76]. The root extract was made with various polar and nonpolar solvents, this all extract was tested on various fungal strains like *C. albicans*, *C. neoformans*, *A. niger*, and *M. audouinii* by using the disc diffusion method and showing potency against the above-mentioned fungal strains. Oil extracted from aerial parts having high antifungal activity against *Candida glabrata*, *Rhizoctonia solani*, *Alternaria solani*, *Fusarium oxysporum*, *Verticillium dahliae*, *Aspergillus nigher*, *Aspergillus flavus*, and *C. albicans* [6].

9.5 CONCLUSIONS

Peppermint (*M. piperita* L.), a popular medicinal herb, is well-known for its flavor and aroma. Products that contain peppermint oil include toothpaste and toothpowder. Mints include high concentrations of isopulegol, menthol, cineole, menthone, and carvone as their main active components. Several species of mint each include a number of unique ingredients, such as glycosides, acylated flavonoids, cinnamic acids, and/or aglycon. Mint is a profitable crop that yields both fresh and dried leaves and essential oil. Herbal teas can be made with these. Essential oils are extracted from the entire plant and used in food goods, cosmetics, nutraceuticals, and pharmaceuticals. When added to food as additives and preservatives, mentha phytoconstituents can help reduce the risk of diabetes. High-fat diets and inactivity are contributing to the global rise in diabetes and cardiovascular disease.

KEYWORDS

- **medicinal properties**
- **peppermint**
- **phenolic compounds**
- **secondary metabolites**

REFERENCES

1. Abdi, G., Karande, V.C., Mohammed, A., Abbasi Tarighat, M., Wen Goh, K., Kari, Z.A., Wei, L.S., Zain, M.R.A.M., Mohammadi, M., Lee, G.E., Barwant, M.M. Pharmacological potential of Sargassum sp. of west coast of Maharashtra Kunkeshwar, India. *Frontiers in Marine Science*, 2022, *9*, 1011218.
2. Adamiec, J., Kalemba, D. Analysis of microencapsulation ability of essential oils during spray drying. *Drying Technology*, 2006, *24*(9), 1127–1132.
3. Agrawal, V.K. *Techniques of Mentha Species Cultivation. Medicinal and Aromatic Plants. Directorate of Extension Services*. IGAU. 2021; pp. 33.
4. Aktsoglou, D.C., Kasampalis, D.S., Sarrou, E., Tsouvaltzis, P., Chatzopoulou, P., Martens, S., Siomos, A.S. Protein hydrolysates supplement in the nutrient solution of soilless grown fresh peppermint and spearmint as a tool for improving product quality. *Agronomy*, 2021, *11*(2), 317. https://doi.org/10.3390/agronomy11020317
5. Akula, R., Ravishankar, G.A. Influence of abiotic stress signals on secondary metabolites in plants. *Plant Signaling & Behavior*, 2011, *6*(11), 1720–1731.
6. Alaklabi, A., Arif, I.A., Ahamed, A., Manilal, A., Surendrakumar, R., Idhayadhulla, A. Larvicidal, nematicidal, antifeedant and antifungal, antioxidant activities of *Mentha spicata* (Lamiaceae) root extracts. *Tropical Journal of Pharmaceutical Research*, 2016, *15*(11), 2383. https://doi.org/10.4314/tjpr.v15i11.12.
7. Alankar, S. A review on peppermint oil. *Asian Journal of Pharmaceutical and Clinical Research*, 2009, *2*(2), 27–33.
8. Alhaithloul, H.A., Soliman, M.H., Ameta, K.L., El-Esawi, M.A., Elkelish, A. Changes in ecophysiology, osmolytes, and secondary metabolites of the medicinal plants of *Mentha piperita* and *Catharanthus roseus* subjected to drought and heat stress. *Biomolecules*, 2019, *10*(1), 43. https://doi.org/10.3390/biom10010043.
9. Almatroodi, S.A., Alsahli, M.A., Almatroudi, A., Khan, A.A., Rahmani, A.H. Peppermint, (mentha × piperita): Role in management of diseases through modulating various biological activities. *Pharmacognosy Journal*, 2021, *13*(3), 822–827. https://doi.org/10.5530/pj.2021.13.104.
10. Al-Sum, B.A., Al-Arfaj, A.A. Antimicrobial activity of the aqueous extract of mint plant. *Science Journal of Clinical Medicine*, 2013, *2*(3), 110–113. https://doi.org/10.11648/j.sjcm.20130203.19.

11. American Diabetes Association. 6. Glycemic targets: Standards of medical care in diabetes—2018. *Diabetes Care*, 2018, *41*(Suppl 1), S55–S64.
12. Anaya, A.L. Allelopathy as a tool in the management of biotic resources in agroecosystems. *Critical Reviews in Plant Sciences*, 1999, *18*(6), 697–739.
13. Areias, F.M., Valentao, P., Andrade, P.B., Ferreres, F., Seabra, R.M. Phenolic fingerprint of peppermint leaves. *Food Chemistry*, 2001, *73*(3), 307–311.
14. Arzani, A., Zeinali, H., Razmjo, K. Iron and magnesium concentrations of mint accessions (Mentha spp.). *Plant Physiology and Biochemistry*, 2007, *45*(5), 323–329.
15. Asao, T., Mochiki, E., Suzuki, H., Nakamura, J.I., Hirayama, I., Morinaga, N., Shoji, H., Shitara, Y., Kuwano, H. An easy method for the intraluminal administration of peppermint oil before colonoscopy and its effectiveness in reducing colonic spasm. *Gastrointestinal Endoscopy*, 2001, *53*(2), 172–177.
16. Aziz, E.E., Gad, N., Khaled, S.M. Effect of cobalt on growth and chemical composition of peppermint plant grown in newly reclaimed soil. *Australian Journal of Basic and Applied Sciences*, 2011, *5*(11), 628–633.
17. Badi, H.N., Yazdani, D., Ali, S.M., Nazari, F. Effects of spacing and harvesting time on herbage yield and quality/quantity of oil in thyme, Thymus vulgaris L. *Industrial Crops and Products*, 2004, *19*(3), 231–236.
18. Barker, S., Grayhem, P., Koon, J., Perkins, J., Whalen, A., Raudenbush, B. Improved performance on clerical tasks associated with administration of peppermint odor. *Perceptual and Motor Skills*, 2003, *97*(3), 1007–1010.
19. Barnard, N.D., Goldman, D.M., Loomis, J.F., Kahleova, H., Levin, S.M., Neabore, S., Batts, T.C. Plant-based diets for cardiovascular safety and performance in endurance sports. *Nutrients*, 2019, *11*(1), 130.
20. Barwant, M.M.M., Manam, D.V.K. *Environment Conservation, Challenges Threats in Conservation of Biodiversity.*
21. Barwant, M.M. Plants as food source; Future prospective. *Local Food and Community Empowerment through Tourism*. 2021; pp. 196–206.
22. Barwant, M.M., Karande, V.C. Review on Nutritive value of Brown seaweed their health benefits. *Journal of the Maharaja Sayajirao University of Baroda ISSN*, 2021, *25*, 0422.
23. Barwant, M.M., Karandeb, V.C. *Seaweed as an Energy Resource: Biomass and Extraction Methods.*
24. Barwant, M., Dawange, A. Preparation of nutritive food supplement from leaf and Pods Moringa oleifera (lamm), Tinospora cordifolia ((Thunb) Miers) and Cinnamomum verum J. presl. *International Journal of Life Sciences*, 47–54.
25. Barwant, M., Lavhate, N. Evaluation of physico-chemical, phytochemical, minerals and antimicrobial properties from selected edible fruits. *International Journal of Research in Chemistry and Environment*, 2020, *10*(1), 1–10.
26. Barwant, M., Ugale, Y., Ghotekar, S., Basnet, P., Nguyen, V.H., Pansambal, S., Murthy, H.C.A., Sillanpaa, M., Bilal, M., Oza, R., Karande, V. Eco-friendly synthesis and characterizations of Ag/AgO/Ag_2O nanoparticles using leaf extracts of *Solanum elaeagnifolium* for antioxidant, anticancer, and DNA cleavage activities. *Chemical Papers*, 2022, *76*(7), 4309–4321.
27. Beigi, M., Torki-Harchegani, M., Ghasemi Pirbalouti, A. Quantity and chemical composition of essential oil of peppermint (Mentha× piperita L.) leaves under

different drying methods. *International Journal of Food Properties*, 2018, *21*(1), 267–276.
28. Benabdallah, A., Boumendjel, M., Aissi, O., Rahmoune, C., Boussaid, M., Messaoud, C. Chemical composition, antioxidant activity and acetylcholinesterase inhibitory of wild Mentha species from northeastern Algeria. *South African Journal of Botany*, 2018, *116*, 131–139.
29. Bijani, A., Cumming, R.G., Hosseini, S.R., Yazdanpour, M., Rahimi, M., Sahebian, A., Ghadimi, R. Obesity paradox on the survival of elderly patients with diabetes: An AHAP-based study. *Journal of Diabetes & Metabolic Disorders*, 2018, *17*, 45–51.
30. Bino, R.J., Hall, R.D., Fiehn, O., Kopka, J., Saito, K., Draper, J., Nikolau, B.J., Mendes, P., Roessner-Tunali, U., Beale, M.H., Trethewey, R.N., Lange, B.M., Wurtele, E.S., Sumner, L.W. Potential of metabolomics as a functional genomics tool. *Trends in Plant Science*, 2004, *9*(9), 418–425. https://doi.org/10.1016/j.tplants.2004.07.004.
31. Borhani Haghighi, A., Motazedian, S., Rezaii, R., Mohammadi, F., Salarian, L., Pourmokhtari, M., Khodaei, S., Vossoughi, M., Miri, R. Cutaneous application of menthol 10% solution as an abortive treatment of migraine without aura: A randomised, double-blind, placebo-controlled, crossed-over study. *International Journal of Clinical Practice*, 2010, *64*(4), 451–456. https://doi.org/10.1111/j.1742-1241.2009.02215.x.
32. Boukhebti, H., Chaker, A.N., Belhadj, H., Sahli, F., Ramdani, M., Laouer, H., Harzallah, D. Chemical composition and antibacterial activity of Mentha pulegium L. and *Mentha spicata* L. essential oils. *Der Pharmacia Lettre*. 2011, *3*(4), 267–275.
33. Brahmi, F., Khodir, M., Mohamed, C., Pierre, D. Chemical composition and biological activities of Mentha species. *Aromatic and Medicinal Plants—Back to Nature*, 2017, *10*, 47–79.
34. Brown, N., John, J.A., Shahidi, F. Polyphenol composition and antioxidant potential of mint leaves. *Food Production, Processing and Nutrition*, 2019, *1*(1), 1. https://doi.org/10.1186/s43014-019-0001-8
35. Buhl, A.E., Waldon, D.J., Baker, C.A., Johnson, G.A. Minoxidil sulfate is the active metabolite that stimulates hair follicles. *Journal of Investigative Dermatology*, 1990, *95*(5), 553–557.
36. Bupesh, G., Amutha, C., Nandagopal, S., Ganeshkumar, A., Sureshkumar, P., Murali, K. Antibacterial activity of *Mentha piperita* L.(peppermint) from leaf extracts-a medicinal plant. *Acta Agriculturae Slovenica*, 2007, *89*(1), 73.
37. Cappellari, L.D.R., Chiappero, J., Palermo, T.B., Giordano, W., Banchio, E. Volatile organic compounds from rhizobacteria increase the biosynthesis of secondary metabolites and improve the antioxidant status in *Mentha piperita* L. grown under salt stress. *Agronomy*, 2020, *10*(8), 1094. https://doi.org/10.3390/agronomy10081094.
38. Caro, D.C., Rivera, D.E., Ocampo, Y., Franco, L.A., Salas, R.D. Pharmacological evaluation of *Mentha spicata* L. and *Plantago major* L., medicinal plants used to treat anxiety and insomnia in Colombian Caribbean coast. *Evidence-Based Complementary and Alternative Medicine*, 2018, *2018*, 5921514.
39. Ćavar Zeljković, S., Schadich, E., Džubák, P., Hajdúch, M., Tarkowski, P. Antiviral activity of selected Lamiaceae essential oils and their monoterpenes against SARS-Cov-2. *Frontiers in Pharmacology*, 2022, *13*, 893634. https://doi.org/10.3389/fphar.2022.893634

40. Chakraborty, K., Chakravarti, A.R., Bhattacharjee, S. Bioactive components of peppermint (*Mentha piperita* L.), their pharmacological and ameliorative potential and ethnomedicinal benefits: A review. *Journal of Pharmacognosy and Phytochemistry*, 2022, *11*(1), 109–114. https://doi.org/10.22271/phyto.2022.v11.i1b.14322
41. Correa, R.C., Peralta, R.M., Haminiuk, C.W., Maciel, G.M., Bracht, A., Ferreira, I.C. New phytochemicals as potential human anti-aging compounds: Reality, promise, and challenges. *Critical Reviews in Food Science and Nutrition*, 2018, *58*(6), 942–957.
42. Cosentino, M., Bombelli, R., Conti, A., Colombo, M.L., Azzetti, A., Bergamaschi, A., Franca, M., Lecchini, S. Antioxidant properties and in vitro immunomodulatory effects of peppermint (*Mentha piperita* l.) essential oils in human leukocytes. *Journal of Pharmaceutical Sciences and Research*, 2009, *1*(3), 33–43.
43. Curutchet, A., Dellacassa, E., Ringuelet, J.A., Chaves, A.R., Viña, S.Z. Nutritional and sensory quality during refrigerated storage of fresh-cut mints (Mentha× piperita and M. spicata). *Food Chemistry*, 2014, *143*, 231–238.
44. Dambrauskienė, E., Viškelis, P., Karklelienė, R. Productivity and biochemical composition of *Mentha piperita* L. of different origin. *Biologija,* 2008, *54*(2).
45. Das, A.K., Nanda, P.K., Bandyopadhyay, S., Banerjee, R., Biswas, S., McClements, D.J. Application of nanoemulsion-based approaches for improving the quality and safety of muscle foods: A comprehensive review. *Comprehensive Reviews in Food Science and Food Safety*, 2020, *19*(5), 2677–2700.
46. Das, G., Patra, J.K., Debnath, T., Ansari, A., Shin, H.S. Investigation of antioxidant, antibacterial, antidiabetic, and cytotoxicity potential of silver nanoparticles synthesized using the outer peel extract of Ananas comosus (L.). *PLoS One*, 2019, *14*(8), e0220950.
47. de Melo, A.N.F., de Souza Pedrosa, G.T., da Cruz Almeida, E.T., Cao, G., Macarisin, D., Schaffner, D.W., Schaffner, D.W., de Souza, E.L., Magnani, M. Successive exposure to *Mentha piperita* L. essential oil affects the culturability and induces membrane repair in a persister epidemic Salmonella typhimurium PT4. *Microbial Pathogenesis*, 2020, *149*, 104264.
48. Derwich, E., Benziane, Z., Taouil, R., Senhaji, O., Touzani, M. Aromatic plants of morocco: GC/MS analysis of the essential oils of leaves of *Mentha piperita*. *Advances in Environmental Biology*, 2010, *4*, 80–85.
49. Dimandja, J.M.D., Stanfill, S.B., Grainger, J., Patterson Jr, D.G. Application of comprehensive two-dimensional gas chromatography (GC× GC) to the qualitative analysis of essential oils. *Journal of High Resolution Chromatography*, 2000, *23*(3), 208–214.
50. Dorman, H.D., Koşar, M., Kahlos, K., Holm, Y., Hiltunen, R. Antioxidant properties and composition of aqueous extracts from Mentha species, hybrids, varieties, and cultivars. *Journal of Agricultural and Food Chemistry*, 2003, *51*(16), 4563–4569.
51. Dutra, K., Wanderley-Teixeira, V., Guedes, C., Cruz, G., Navarro, D., Monteiro, A., Agra, A., Neto, C.L., Teixeira, Á. Toxicity of essential oils of leaves of plants from the genus Piper with influence on the nutritional parameters of Spodoptera frugiperda (JE Smith) (Lepidoptera: Noctuidae). *Journal of Essential Oil Bearing Plants*, 2020, *23*(2), 213–229.
52. Eccles, R. Menthol: effects on nasal sensation of airflow and the drive to breathe. *Current Allergy and Asthma Reports*, 2003, *3*(3), 210–214.

53. Edris, A.E. Pharmaceutical and therapeutic potentials of essential oils and their individual volatile constituents: A review. *Phytotherapy Research: An International Journal Devoted to Pharmacological and Toxicological Evaluation of Natural Product Derivatives*, 2007, 21(4), 308–323.
54. Eftekhari, A., Khusro, A., Ahmadian, E., Dizaj, S.M., Hasanzadeh, A., Cucchiarini, M. Phytochemical and nutra-pharmaceutical attributes of Mentha spp.: A comprehensive review. *Arabian Journal of Chemistry*, 2021, *14*(5), 103106.
55. Farag, M.A., Abib, B., Ayad, L., Khattab, A.R. Sweet and bitter oranges: An updated comparative review of their bioactives, nutrition, food quality, therapeutic merits and biowaste valorization practices. *Food Chemistry*, 2020, *331*, 127306.
56. Fonnegra, F.G. *Plantas Medicinales Aprobadas en Colombia*. Universidad de Antioquia. 2007.
57. Franciscato, L.M.S.S., Silva, M.R., Andrich, F., Sakai, O.A., Doyama, J.T., Júnior, A. F., Mortinz, C.M.F. Chemical characterization and antimicrobial activity of essential oils of mint (*Mentha spicata* L.) and surinam cherry (*Eugenia uniflora* L.). *Orbital: The Electronic Journal of Chemistry*, 2018, *10*(6). https://doi.org/10.17807/orbital.v10i6.1008.
58. Franco, R.R., Alves, V.H.M., Zabisky, L.F.R., Justino, A.B., Martins, M.M., Saraiva, A.L., Goulart, L.R., Espindola, F.S. Antidiabetic potential of Bauhinia forficata Link leaves: a non-cytotoxic source of lipase and glycoside hydrolases inhibitors and molecules with antioxidant and antiglycation properties. *Biomedicine & Pharmacotherapy*, 2020, *123*, 109798.
59. Garcia-Mier, L., Meneses-Reyes, A.E., Jimenez-Garcia, S.N., Mercado Luna, A., García Trejo, J.F., Contreras-Medina, L.M., Feregrino-Perez, A.A. Polyphenol content and antioxidant activity of stevia and peppermint as a result of organic and conventional fertilization. *Journal of Food Quality*, 2021, 1–6. https://doi.org/10.1155/2021/6620446.
60. Gatti, R., Lotti, C., Morigi, R., Andreani, A. Determination of octopamine and tyramine traces in dietary supplements and phytoextracts by high performance liquid chromatography after derivatization with 2,5-dimethyl-1H-pyrrole-3, 4-dicarbaldehyde. *Journal of Chromatography A*, 2012, *1220*, 92–100.
61. Getahun, Z., Asres, K., Mazumder, A. Essential oil composition, antibacterial and antioxidant activities of four mentha species growing in Ethiopia. *Ethiopian Pharmaceutical Journal*, 2008, *25*(2), 9–16. https://doi.org/10.4314/epj.v25i2.35123.
62. Geuenich, S., Goffinet, C., Venzke, S., Nolkemper, S., Baumann, I., Plinkert, P., Reichling, J., Keppler, O.T. Aqueous extracts from peppermint, sage and lemon balm leaves display potent anti-HIV-1 activity by increasing the virion density. *Retrovirology*, 2008, *5*, 27. https://doi.org/10.1186/1742-4690-5-27.
63. Ghanemi, A., Melouane, A., Yoshioka, M., St-Amand, J. Exercise and high-fat diet in obesity: Functional genomics perspectives of two energy homeostasis pillars. *Genes*, 2020, *11*(8), 875.
64. Gholamipourfard, K., Salehi, M., Banchio, E. *Mentha piperita* phytochemicals in agriculture, food industry and medicine: Features and applications. *South African Journal of Botany*, 2021, *141*, 183–195.

65. Gobel, H., Schmidt, G., Soyka, D. Effect of peppermint and eucalyptus oil preparations on neurophysiological and experimental algesimetric headache parameters. *Cephalalgia*, 1994, *14*(3), 228–234.
66. Goerg, K.J., Spilker, T.H. Effect of peppermint oil and caraway oil on gastrointestinal motility in healthy volunteers: A pharmacodynamic study using simultaneous determination of gastric and gall-bladder emptying and orocaecal transit time. *Alimentary Pharmacology & Therapeutics*, 2003, *17*(3), 445–451. https://doi.org/10.1046/j.0269-2813.2003.01421.x.
67. Hever, J. Plant-based diets: A physician's guide. *The Permanente Journal*, 2016, *20*(3), 15–82.
68. Hopkins, B.D., Goncalves, M.D., Cantley, L.C. Insulin–PI3K signalling: an evolutionarily insulated metabolic driver of cancer. *Nature Reviews Endocrinology*, 2020, *16*(5), 276–283.
69. HOSEIN, Z.M., Kiarostami, K.H., ILKHANI, Z.M., Saboora, A. A study on allelopathic compounds derived from Hordeum spontaneum on carbohydrates, proteins and some enzymes of wheat (*Triticum aestivum* L.). 2009.
70. Hussain, A.I., Anwar, F., Shahid, M., Ashraf, M., Przybylski, R. Chemical composition, and antioxidant and antimicrobial activities of essential oil of spearmint (*Mentha spicata* L.) from Pakistan. *Journal of Essential Oil Research*, 2010, *22*(1), 78–84. https://doi.org/10.1080/10412905.2010.9700269.
71. İşcan, G., Kirimer, N., Kürkcüoğlu, M., Başer, H.C., Demirci, F. Antimicrobial screening of *Mentha piperita* essential oils. *Journal of Agricultural and Food Chemistry*, 2002, *50*(14), 3943–3946.
72. Jones, W.P., Chin, Y.W., Kinghorn, A.D. The role of pharmacognosy in modern medicine and pharmacy. *Current Drug Targets*, 2006, *7*(3), 247–264.
73. Kapp, K. Polyphenolic and essential oil composition of Mentha and their antimicrobial effect. Dissertations, University of Helsinki, 2015.
74. Kapp, K., Hakala, E., Orav, A., Pohjala, L., Vuorela, P., Püssa, T., Vuorela, H., Raal, A. Commercial peppermint (Mentha× piperita L.) teas: Antichlamydial effect and polyphenolic composition. *Food Research International*, 2013, *53*(2), 758–766.
75. Katariya, C., Malaiappan, S. Antimicrobial activity of mint extract on periodontopathic bacteria—An in-vitro study. *Journal of Research in Medical and Dental Science*, 2022, 10(1), 509–512.
76. Kedia, A., Prakash, B., Mishra, P.K., Chanotiya, C.S., Dubey, N.K. Antifungal, antiaflatoxigenic, and insecticidal efficacy of spearmint (*Mentha spicata* L.) essential oil. *International Biodeterioration and Biodegradation*, 2014, *89*, 29–36. https://doi.org/10.1016/j.ibiod.2013.10.027.
77. Khorasaninejad, S., Mousavi, A., Soltanloo, H., Hemmati, K., Khalighi, A. The effect of drought stress on growth parameters, essential oil yield and constituent of Peppermint (*Mentha piperita* L.). *Journal of Medicinal Plants Research*, 2011, *5*(22), 5360–5365. http://www.academicjournals.org/JMPR.
78. Kolhe, S., Barwant, M. Comparative study of biofertilizers and chemical fungicides on Zea Mays and Glycine Max. *International Journal of Innovative Science & Technology*, 2021, *6*(2), 72–79.

79. Kumar, K.V., Patra, D.D. Alteration in yield and chemical composition of essential oil of *Mentha piperita* L. plant: Effect of fly ash amendments and organic wastes. *Ecological Engineering*, 2012, *47*, 237–241.
80. Lange, B.M., Mahmoud, S.S., Wildung, M.R., Turner, G.W., Davis, E.M., Lange, I., Baker, R.C., Boydston, R.A., Croteau, R.B. Improving peppermint essential oil yield and composition by metabolic engineering. *Proceedings of the National Academy of Sciences*, 2011, *108*(41), 16944–16949.
81. Lavhate, N., Barwant, M. nutritional quality of different nuts as revealed by its biochemical analysis. *International Journal of Research in Applied, Natural and Social Sciences*, 8(4), 1–12.
82. Li, Y., Liu, Y., Ma, A., Bao, Y., Wang, M., Sun, Z. In vitro antiviral, anti-inflammatory, and antioxidant activities of the ethanol extract of *Mentha piperita* L. *Food Science and Biotechnology*, 2017, *26*(6), 1675–1683. https://doi.org/10.1007/s10068-017-0217-9.
83. Liu, K.H., Zhu, Q., Zhang, J.J., Xu, J.F., Wang, X.C. Chemical composition and biological activities of the essential oil of *Mentha spicata* Lamiaceae. *Advanced Materials Research*, 2012, *524–527*, 2269–2272. https://doi.org/10.4028/www.scientific.net/AMR.524-527.2269.
84. Loolaie, M., Moasefi, N., Rasouli, H., Adibi, H. Peppermint and its functionality: A review. *Archives of Clinical Microbiology*, 2017, *8*(4), 54.
85. López, V., Martín, S., Gómez-Serranillos, M.P., Carretero, M.E., Jäger, A.K., Calvo, M.I. Neuroprotective and neurochemical properties of mint extracts. *Phytotherapy Research*, 2010, *24*(6), 869–874. https://doi.org/10.1002/ptr.3037, PubMed: 19943334.
86. Maffei, M. Sustainable methods for a sustainable production of peppermint (Mentha x piperita L.) essential oil. *Journal of Essential Oil Research*, 1999, *11*(3), 267–282.
87. Mahdavikia, F., Saharkhiz, M.J. Secondary metabolites of peppermint change the morphophysiological and biochemical characteristics of tomato. *Biocatalysis and Agricultural Biotechnology*, 2016, *7*, 127–133. http://dx.doi.org/10.1016/j.bcab.2016.05.013.
88. Mahdavikian, S., Fallahi, M., Khatony, A. Comparing the effect of aromatherapy with peppermint and lavender essential oils on fatigue of cardiac patients: A randomized controlled trial. *Evidence-Based Complementary and Alternative Medicine*, 2021, *2021*, 9925945. https://doi.org/10.1155/2021/9925945.
89. Mahendran, G., Rahman, L.U. Ethnomedicinal, phytochemical and pharmacological updates on Peppermint (Mentha× piperita L.)—A review. *Phytotherapy Research*, 2020, *34*(9), 2088–2139. https://doi.org/10.1002/ptr.6664.
90. Mahmoudi, S., Nasiri, R., Jafari Sales, A. In-vitro antibacterial effects of methanolic extract of peppermint (*Mentha piperita* Lamiaceae) on standard Staphylococcus aureus, Bacillus cereus, Escherichia coli and Pseudomonas aeruginosa strain. *Jorjani Biomedicine Journal*, 2019, *7*(4), 4–10. https://doi.org/10.29252/jorjanibiomedj.7.4.4.
91. Mainasara, M.M., Bakar, M.F.A., Waziri, A.H., Musa, A.R. (2018). Comparison of phytochemical, proximate and mineral composition of fresh and dried peppermint (*Mentha piperita*) leaves. *Journal of Science and Technology*, 10(2). https://doi.org/10.30880/jst.2018.10.02.014.

92. Mauskop, A. Alternative therapies in headache: Is there a role? *Medical Clinics of North America*, 2001, *85*(4), 1077–1084.
93. McConkey, M.E., Gershenzon, J., Croteau, R.B. Developmental regulation of monoterpene biosynthesis in the glandular trichomes of peppermint. *Plant Physiology*, 2000, *122*(1), 215–224.
94. McMurry, J.E. *Organic Chemistry with Biological Applications*. Cengage Learning. 2014.
95. Mench, M., Schwitzguébel, J.P., Schroeder, P., Bert, V., Gawronski, S., Gupta, S. Assessment of successful experiments and limitations of phytotechnologies: Contaminant uptake, detoxification and sequestration, and consequences for food safety. *Environmental Science and Pollution Research*, 2009, *16*, 876–900.
96. Messenger, A.G., Rundegren, J. Minoxidil: Mechanisms of action on hair growth. *British Journal of Dermatology*, 2004, *150*(2), 186–194.
97. Mirhendi, H., Makimura, K., Khoramizadeh, M., Yamaguchi, H. A one-enzyme PCR-RFLP assay for identification of six medically important Candida species. *Japanese Journal of Medical Mycology*, 2006, *47*(3), 225–229. https://doi.org/10.3314/jjmm.47.225.
98. Mirhendi, H., Makimura, K., Zomorodian, K., Maeda, N., Ohshima, T., Yamaguchi, H. Differentiation of Candida albicans and Candida dubliniensis using a single-enzyme PCR-RFLP method. *Japanese Journal of Infectious Diseases*, 2005, *58*(4), 235–237.
99. Moghaddam, M., Pourbaige, M., Tabar, H.K., Farhadi, N., Ahmadi Hosseini, S.M.A. Composition and antifungal activity of peppermint (*Mentha piperita*) essential oil from Iran. *Journal of Essential Oil Bearing Plants*, 2013, *16*(4), 506–512. https://doi.org/10.1080/0972060X.2013.813265.
100. Mohan, K., Ganesan, A.R., Muralisankar, T., Jayakumar, R., Sathishkumar, P., Uthayakumar, V., Revathi, N. Recent insights into the extraction, characterization, and bioactivities of chitin and chitosan from insects. *Trends in Food Science & Technology*, 2020, *105*, 17–42.
101. Naeem, M., Muqarab, R., Waseem, M. The Solanum melongena COP1 delays fruit ripening and influences ethylene signaling in tomato. *Journal of Plant Physiology*, 2019, *240*, 152997.
102. Nair, B. Final report on the safety assessment of Mentha Piperita (Peppermint) Oil, Mentha Piperita (Peppermint) Leaf Extract, Mentha Piperita (Peppermint) Leaf, and Mentha Piperita (Peppermint) leaf water. *International Journal of Toxicology*, 2001, *20*, 61–73.
103. Ng, K.R., Lyu, X., Mark, R., Chen, W.N. Antimicrobial and antioxidant activities of phenolic metabolites from flavonoid-producing yeast: Potential as natural food preservatives. *Food Chemistry*, 2019, *270*, 123–129.
104. Nickavar, B., Alinaghi, A., Kamalinejad, M. Evaluation of the antioxidant properties of five Mentha species. *Iranian Journal of Pharmaceutical Research*, 2008, *7*(3), 203–209.
105. Niggli, U., Slabe, A., Schmid, O., Halberg, N., Schlüter, M. Vision for an organic food and farming research agenda 2025. *Organic Knowledge for the Future*, 2008.
106. Ojewumi, M.E., Adedokun, S.O., Omodara, O.J., Oyeniyi, E.A., Taiwo, O.S., Ojewumi, E.O. Phytochemical and antimicrobial activities of the leaf oil extract

of *Mentha spicata* and its efficacy in repelling mosquito. *International Journal of Pharmaceutical Research & Allied Sciences*, 2017, *6*(4), 17–27.

107. Ommi, D., Saeedi, N., Hajizadeh, N., Mohseni, G., Teymourian, H. Complementary effects of *Mentha piperita* (Peppermint) and Rosa damascene extract (Rose oil) on SpO2 in patients with COVID-19: A randomized clinical trial. *Herbal Medicines Journal (Herb Med J)*, 2020, *5*(3), 119–126.
108. Padmini, E., Prema, K., Vijaya Geetha, B., Usha Rani, M. Comparative study on composition and antioxidant properties of mint and black tea extract. *International Journal of Food Science & Technology*, 2008, *43*(10), 1887–1895.
109. Patel, T., Ishiuji, Y., Yosipovitch, G. Menthol: A refreshing look at this ancient compound. *Journal of the American Academy of Dermatology*, 2007, *57*(5), 873–878.
110. Pereira, E., Pimenta, A.I., Calhelha, R.C., Antonio, A.L., Verde, S.C., Barros, L., ... Ferreira, I.C. Effects of gamma irradiation on cytotoxicity and phenolic compounds of Thymus vulgaris L. and Mentha x piperita L. *LWT-Food Science and Technology*, 2016, *71*, 370–377.
111. Pérez, M.G.F., Rocha-Guzmán, N.E., Mercado-Silva, E., Loarca-Piña, G., Reynoso-Camacho, R. Effect of chemical elicitors on peppermint (*Mentha piperita*) plants and their impact on the metabolite profile and antioxidant capacity of resulting infusions. *Food Chemistry*, 2014, *156*, 273–278.http://dx.doi.org/10.1016/j.foodchem.2014.01.101.
112. Pipitone, G., Zoppi, G., Bocchini, S., Rizzo, A.M., Chiaramonti, D., Pirone, R., Bensaid, S. Aqueous phase reforming of the residual waters derived from lignin-rich hydrothermal liquefaction: Investigation of representative organic compounds and actual biorefinery streams. *Catalysis Today*, 2020, *345*, 237–250.
113. Qamar, N., Pandey, M., Vasudevan, M., Kumar, A., Shasany, A.K. Glandular trichome specificity of menthol biosynthesis pathway gene promoters from Mentha× piperita. *Planta*, 2022, *256*(6), 110.
114. Quintin, D., Garcia-Gomez, P., Ayuso, M., Sanmartin, A.M. Active biocompounds to improve food nutritional value. *Trends in Food Science & Technology*, 2019, *84*, 19–21.
115. Raju, M., Varakumar, S., Lakshminarayana, R., Krishnakantha, T.P., Baskaran, V. Carotenoid composition and vitamin A activity of medicinally important green leafy vegetables. *Food Chemistry*, 2007, *101*(4), 1598–1605.
116. Raudenbush, B., Koon, J., Meyer, B., Flower, N. Effects of ambient odor on pain threshold, pain tolerance, mood, workload, and anxiety. In *Second Annual Meeting of the Society for Psychophysiological Research*. Washington DC: Society for Psychophysiological Research. 2002.
117. Raudenbush, B., Meyer, B., Eppich, B. The effects of odors on objective and subjective measures of athletic performance. *International Sports Journal*, 2002, *6*(1), 14–27.
118. Reddy, D.N., Al-Rajab, A.J., Sharma, M., Moses, M.M., Reddy, G.R., Albratty, M. Chemical constituents, in vitro antibacterial and antifungal activity of Mentha × piperita L. (peppermint) essential oils. *Journal of King Saud University—Science*, 2019, *31*(4), 528–533.
119. Riachi, L.G., De Maria, C.A. Peppermint antioxidants revisited. *Food Chemistry*, 2015, *176*, 72–81. https://doi.org/10.1016/j.foodchem.2014.12.028.

120. Ritota, M., Manzi, P. Natural preservatives from plant in cheese making. *Animals*, 2020, *10*(4), 749.
121. Saharkhiz, M.J., Smaeili, S., Merikhi, M. Essential oil analysis and phytotoxic activity of two ecotypes of Zataria multiflora Boiss. growing in Iran. *Natural Product Research*, 2010, *24*(17), 1598–1609.
122. Saqib, S., Ullah, F., Naeem, M., Younas, M., Ayaz, A., Ali, S., Zaman, W. Mentha: Nutritional and health attributes to treat various ailments including cardiovascular diseases. *Molecules*, 2022, *27*(19), 6728. https://doi.org/10.3390/molecules27196728.
123. Satmi, F.R.S., Hossain, M.A. In vitro antimicrobial potential of crude extracts and chemical compositions of essential oils of leaves of *Mentha piperita* L native to the Sultanate of Oman. *Pacific Science Review A: Natural Science and Engineering*, 2016, *18*(2), 103–106.
124. Schoenfeld, B. Effects of oral supplementation of mint extract on muscle pain and blood lactate. *Biomedical Human Kinetics*, 2010, *2*(2010), 66–69.
125. Schuhmacher, A., Reichling, J., Schnitzler, P. Virucidal effect of peppermint oil on the enveloped viruses herpes simplex virus type 1 and type 2 in vitro. *Phytomedicine: International Journal of Phytotherapy and Phytopharmacology*, 2003, *10*(6–7), 504–510. https://doi.org/10.1078/094471103322331467.
126. Serlahwaty, D., Giovani, C. In silico screening of mint leaves compound (*Mentha piperita* L.) as a potential inhibitor of SARS-CoV-2. *Pharmacy Education*, 2021, *21*(2), 81–86. https://doi.org/10.46542/pe.2021.212.8186.
127. Shapiro, J., Price, V.H. Hair regrowth: Therapeutic agents. *Dermatologic Clinics*, 1998, *16*(2), 341–356.
128. Shrivastava, A. A review on peppermint oil. *Asian Journal of Pharmaceutical and Clinical Research*, 2009, *2*(2), 27–33.
129. Singh, J.P.V., Selvendiran, K., Banu, S.M., Padmavathi, R., Sakthisekaran, D. Protective role of Apigenin on the status of lipid peroxidation and antioxidant defense against hepatocarcinogenesis in Wistar albino rats. *Phytomedicine*, 2004, *11*(4), 309–314. https://doi.org/10.1078/0944711041495254.
130. Singh, R., Shushni, M.A., Belkheir, A. Antibacterial and antioxidant activities of *Mentha piperita* L. *Arabian Journal of Chemistry*, 2015, *8*(3), 322–328. http://doi.org/10.1016/j.arabjc.2011.01.019.
131. Soares, A.R., de Cássia Siqueira-Soares, R., Salvador, V.H., de Lourdes Lucio Ferrarese, M., Ferrarese-Filho, O. The effects of L-DOPA on root growth, lignification and enzyme activity in soybean seedlings. *Acta Physiologiae Plantarum*, 2012, *34*, 1811–1817.
132. Spirling, L.I., Daniels, I.R. Botanical perspectives on health peppermint: more than just an after-dinner mint. *The Journal of the Royal Society for the Promotion of Health*, 2001, *121*(1), 62–63.
133. Stamp, N. Out of the quagmire of plant defense hypotheses. *The Quarterly Review of Biology*, 2003, *78*(1), 23–55.
134. Stanev, S., Zheljazkov, V.D. Study on Essential oil and Free Menthol Accumulation in 19 cultivars, populations, and clones of peppermint (Mentha X piperita). In: *XXVI International Horticultural Congress: The Future for Medicinal and Aromatic Plants 629*. 2002; pp. 149–152.

135. Stough, D., Stenn, K., Haber, R., Parsley, W.M., Vogel, J.E., Whiting, D.A., Washenik, K. Psychological effect, pathophysiology, and management of androgenetic alopecia in men. *Mayo Clinic Proceedings*, 2005, *80*(10), 1316–1322.
136. Sun, Z., Wang, H., Wang, J., Zhou, L., Yang, P. Chemical composition and anti-inflammatory, cytotoxic and antioxidant activities of essential oil from leaves of *Mentha piperita* grown in China. *PLoS One*, 2014, *9*(12), e114767. http://doi.org/10.1371/journal.pone.0114767.
137. Taherpour, A.A., Khaef, S., Yari, A., Nikeafshar, S., Fathi, M., Ghambari, S. Chemical composition analysis of the essential oil of *Mentha piperita* L. from Kermanshah, Iran by hydrodistillation and HS/SPME methods. *Journal of Analytical Science and Technology*, 2017, *8*, 1–6. http://doi.org/10.1186/s40543-017-0122-0.
138. Torres, A., Oliva, R.M., Castellano, D., Cross, P. *First World Congress on Allelopathy—A Science of the Future*. Cadiz, Spain. 1996.
139. Trevisan, S.C.C., Menezes, A.P.P., Barbalho, S.M., Guiguer, É.L. Properties of mentha piperita: a brief review. *World Journal of Pharmaceutical and Medical Research*, 2017, *3*(1), 309–313.
140. Triantaphyllou, K., Blekas, G., Boskou, D. Antioxidative properties of water extracts obtained from herbs of the species Lamiaceae. *International Journal of Food Sciences and Nutrition*, 2001, *52*(4), 313–317.
141. Trüeb, R.M. Molecular mechanisms of androgenetic alopecia. *Experimental Gerontology*, 2002, *37*(8–9), 981–990.
142. Tsai, M.L., Wu, C.T., Lin, T.F., Lin, W.C., Huang, Y.C., Yang, C.H. Chemical composition and biological properties of essential oils of two mint species. *Tropical Journal of Pharmaceutical Research*, 2013, *12*(4), 577–582.
143. Tyagi, A.K., Malik, A. Antimicrobial potential and chemical composition of *Mentha piperita* oil in liquid and vapour phase against food spoiling microorganisms. *Food Control*, 2011, *22*(11), 1707–1714.
144. Umezu, T., Sakata, A., Ito, H. Ambulation-promoting effect of peppermint oil and identification of its active constituents. *Pharmacology Biochemistry and Behavior*, 2001, *69*(3–4), 383–390.
145. Uribe, E., Marín, D., Vega-Gálvez, A., Quispe-Fuentes, I., Rodríguez, A. Assessment of vacuum-dried peppermint (*Mentha piperita* L.) as a source of natural antioxidants. *Food Chemistry*, 2016, *190*, 559–565.
146. Voirin, B., Bayet, C., Faure, O., Jullien, F. Free flavonoid aglycones as markers of parentage in Mentha aquatica, M. citrata, M. spicata and M. x piperita. *Phytochemistry*, 1999, *50*(7), 1189–1193.
147. Westman, E.C., Tondt, J., Maguire, E., Yancy Jr, W.S. Implementing a low-carbohydrate, ketogenic diet to manage type 2 diabetes mellitus. *Expert Review of Endocrinology & Metabolism*, 2018, *13*(5), 263–272.
148. Yang, L., Wen, K.S., Ruan, X., Zhao, Y.X., Wei, F., Wang, Q. Response of plant secondary metabolites to environmental factors. *Molecules*, 2018, *23*(4), 762.
149. Yucharoen, R., Meepowpan, P., Tragoolpua, Y. Inhibitory effect of peppermint extracts and menthol against Herpes Simplex Virus infection. *Chiang Mai Journal of Science*, 2012, *39*(1), 97–110. http://it.science.cmu.ac.th/ejournal.

150. Zheljazkov, V., Margina, A. Effect of increasing doses of fertilizer application on quantitative and qualitative characters of mint. *Acta Horticulturae*, 1995, *426*, 579–592.

151. Zheng, W., Wang, S.Y. Antioxidant activity and phenolic compounds in selected herbs. *Journal of Agricultural and Food Chemistry*, 2001, *49*(11), 5165–5170.

152. Zoladz, P., Raudenbush, B., Lilley, S. Cinnamon perks performance. In: *April 21-25 Proceedings of Annual Meeting of the Association for Chemoreception Sciences*, Sarasota, FL, USA. 2004.

CHAPTER 10

Cropping System in Peppermint

DHARINI CHITTARAGI,[1*] LOKESH S.,[2] and R. SURESH[3]

[1]*Department of Plantation, Spices, Medicinal and Aromatic Crops, Kittur Rani Channamma College of Horticulture, Arabhavi, University of Horticultural Sciences, Bagalkot, Karnataka, India*

[2]*Department of Agricultural Economics, Post-Graduate College of Agriculture, RPCAU, Pusa, Bihar*

[3]*Directorate of Agribusiness Development, TNAU, Coimbatore, Tamil Nadu, India*

*Corresponding author

ABSTRACT

India is the world leader in mint oil production. It is a perennial herb that belongs to the family Lamiaceae, which includes various botanical forms like Japanese, pepper, spearmint, garden, and orange mint that yield essential oil on distillation. These are mainly used in cosmetics, pharmaceuticals, food, and confectionery. Cropping systems followed in the cultivation of the pepper mint influence the fresh leaf yield and oil percentage. Different cropping systems, such as intercropping (sugarcane, rye, faba bean, and soya), relay cropping (wheat), crop rotation (mint–maize–potato, mint early paddy potato), showed positive effects with pepper mint. Interactions between these cropping patterns would influence the growth characteristics of the companion crop. They will also enhance the higher food production, ensuring the food security of the people in rural and urban areas, and provide a new era of jobs, helping to achieve sustainable farm income in the future

Peppermint: Novel Strategies for Performance Improvement. Bilal Amhad, Tariq Ahmad Dar, and Khalid Rehman Hakeem (Eds.)

DOI: 10.1201/9781779641557-10

10.1 INTRODUCTION

Mint is an aromatic plant that is mainly propagated by using root suckers. It is perennial in nature. Botanically, different types of mints exist, namely, Japanese/corn/field (*Mentha arvensis*), peppermint (*Mentha piperita* L.), spear/garden/lamb (*Mentha spicata* L.), and Bergamot mint or orange mint (*Mentha citrata* Ehrh.) belong to the family Lamiaceae; Labiatae. The mint oil is obtained from *M. arvensis*, *M. piperita*, and *M. spicata* on distillation. These are industrially used in cosmetics, pharmaceuticals, food, and confectionery.

The origin of mint is mainly in the European region. Peppermint is grown in the USA, Morocco, Argentina, Australia, France, USSR, Bulgaria, Czechoslovakia, Hungary, Italy, Switzerland, etc. The highest production of mint oil is in the USA.

M. piperita is generally grown in humid and temperate conditions like Kashmir and the hills of Uttar Pradesh and Himachal Pradesh. Generally grows to a height of around 45–90 cm. The fresh herb contains essential oils ranging from 0.4 to 0.6%. The constituents of peppermint oil are terpenes, menthyl acetate (14–15%), menthone (9–25%), and menthofuran.

M. piperita is found to grow well in various soil and climatic conditions as compared to other species [15]. Peppermint oil is one of the most popular and widely used essential oils. It is used in the preparation of chewing gums, confectionery, and alcoholic liquors. The mint oil has 50% menthol, which is the highest among all other species. Mint oil and menthol are exported, and there is scope for producing l-limonene, 3-octanol, pure menthone, and isomenthone.

10.2 FARMING SYSTEMS

This is a composed system that includes all the farm practices such as cultivation of various crops, rearing of cattle, fishes, poultry, growing of trees by using all the resources to produce food for their living and earning profits with social well-being.

10.3 CROPPING SYSTEMS

It is a system of growing different crops in a sequential manner on the same piece of land within a given time that interacts with the surrounding environment.

10.3.1 SCOPE OF CROPPING SYSTEM

Increased total food production per land unit per year, together contributes for increased farmer's income as follows:

- Sustaining the soil health;
- Improved distribution of income throughout the year;
- Increased total employment of labor throughout the year; and
- Improved nutrition for the farm family from crop diversification.

This is the use of a cropping system as a strategy for increasing productivity and the income of small farmers

10.3.2 RESOURCE MANAGEMENT IN CROPPING SYSTEM

Cropping system implies an association of different crops and their interaction. An interaction is essentially a response of one species to the environment as modified by the presence of another species. Such interactions are the characteristics of any system, and they greatly influence the management of the system. The interaction in sequential cropping differs in yield and quality as compared to other systems.

This includes only the crops that belong to similar species only, whereas, intercropping includes the crops that belong to other species.

The possible effects of such interactions can be the following:

- Difference in spatial and time dimensions in the crop associations.
- Complementary effects of crop associations.
- Competition for available resources like light, water, and nutrients.
- Effect on soil fertility and nutrient availability, and uptake.
- Possible changes in pests, disease, and weed incidence.
- Adverse effects due to allelopathy.

Management of cropping systems must aim at increasing the complementary interactions and minimizing the competitive interactions between the component crops so that the overall productivity of the system does not suffer.

Cropping system management has to be studied under the following aspects or situations:

- Selection of crops and varieties
- Tillage and land shaping

- Plant populations, crop geometry, method, and time of planting
- Water management
- Weed management
- Soil fertility management and fertilizer use
- Plant protection measures
- Labor management

Other management practices for specific objectives are as follows: Select different crops and specific varieties for sequential cropping. Irrigation water availability was adopted.

Soil type: Fertile soil (6–7.5 pH).

Profit: Based on the price variation and proximity to the town.

Multiple requirements of farm and farmer: Farmer's family/ livestock requirements for his daily life. (The best system will have 60–70% area with food (30%) and fodder (10%) crops.)

Maintaining soil fertility: The nature of the crops selected in the intensive cropping should help to maintain the soil fertility and not exhaust it completely, including a legume in the cropping to maintain the fertility of the soil.

Crops which may serve as alternate hosts: For certain pests and diseases, alternate host crops must be avoided as succeeding crops.

Crops that make the best utilization of radiations: Crops that make the use of radiations included to the extent possible. For example, C4 plants like maize, sugarcane, etc. must be included, which have rapid canopy development in summer when the solar radiation is high.

Yield: Short- and medium-duration types (photo-insensitive) can be recommended for all seasons and yield higher productivity.

Leaf area: In an intercropping system, due to the shading effect of tall trees, the short plants will receive filtered sunlight, which will help to improve their photosynthetic activity due to increased leaf area, weight, and ratio.

Leaf inclination: This is the position of the leaf in plants. For an intercropping system, tall trees must have an erect type and smaller one's must be of a horizontal type, which will help for proper absorption of sunlight.

Growth duration: The two crops' duration varies and also their requirement of light for proper growth.

Competition for nutrients: This is mainly based on the crop's nature, quantity of the different nutrients required for the physiological and biochemical activities of each crop.

Root distribution pattern: The root system of one crop overlaps with the other crop, making the competition among them for the nutrients. Hence, one should select the crop for intercropping with a varied type of rooting pattern for both the crops.

Crop duration: This must vary for all the crops, including the main crop, as it will create competition among them, which will affect their growth, physiology, and nutrient uptake.

10.3.2.1 ADVANTAGES

- Better utilization of natural resources and applied nutrients.
- Improvement of soil fertility.
- Reduces the risk of soil erosion.
- Better control of weeds.
- Increases productivity.
- Provides additional employment for the family.
- Increases the income of the farmer.

10.3.2.2 SOCIAL AND ECONOMIC ADVANTAGE

- Avoids dependence on only one crop.
- Reduction in the excess use of fertilizers.
- Possible to recover investments in a very short time.
- Employment generation.
- The farmer with less economic resources can produce a large variety of useful products.
- Components can constitute a type of "savings" for the future.
- Efficient use of dry lands and marginal lands.

10.4 CROP ROTATION

The crop rotations followed in Uttar Pradesh are mint–maize–potato ; mint-early paddy–potato; mint-late paddy–sweet pea. In case of Punjab, mint–maize–rapeseed (or mustard) , mint–paddy and mint–maize–potato are widely practiced.

10.5 INTERCROPPING

The intercropping will help for the proper usage of available land and resources [7]. It helps for higher production [10, 16].

Mint is supposed to be the best crop for intercropping due to its quick growth, higher yield within less time period. *M. piperita* has a wide adaptation ability in different climate and soil conditions, allowing to achieve a higher quality of oil [15]. The yield of peppermint essential oil is not only controlled by genes but also influenced by climatic variations (temperature, rainfall, day length, and radiation) and nutrient uptake [2].

10.5.1 INTERCROP OF SUGARCANE AND MINT

Among the intercropping systems, menthol mint and sugarcane (2 rows: 180 cm) showed better results for the growth parameters and yield [1].

Experiments conducted on intercropping of mint with maize rape and onion at Ludhiana showed maximum fresh leaf yield and oil content in sole mint followed by mint and onion [3].

Different Mentha species intercropped in a poplar agroforestry system showed better performance of herbage and essential oil yield [6].

Intercropping showed the higher levels of menthol production along with mint with faba bean [12].

Intercropping may also have implications for insect control and irrigation requirements. An intercropping experiment was conducted to analyze the influence of peppermint on the pest' (*Drosophila suzukii*) population in the fruit orchards. The mint produced high levels of volatile oil compounds that may help to reduce the pest population [4].

Experiments with rye and mint intercropping showed better performance and recorded higher yield during early spring may be due to lower weed population and reduced nitrate leaching [13].

Higher fresh yield was observed when the mint was intercropped with carrot and lettuce [5].

The three different species of mint intercropping caused a reduction in sugarcane yield to the extent of 17.8, 18.1, and 30.7%, respectively [8]. Randhawa et al. [14] observed that the cane yield was lower when *Mentha*, an essential oil crop, was grown as an intercrop with sugarcane.

Soyabean intercropped with mint recorded higher levels of leaf yield and essential oil with good quality (menthol) in monoculture, followed by 3:2 and 3:1 ratios [11].

10.5.2 RELAY CROPPING

The relay cropping of bread wheat and mint showed higher economic returns [9]. Similarly, the mint planted *via* suckers recorded 32% extra income over wheat.

10.6 CONCLUSION

Cropping systems help to produce more food, which helps for social well-being of the community and also increases job vacancies for youngsters, which will reduce poverty level with enhanced income.

It is an important agribusiness way to create more wealth and capital resources.

KEYWORDS

- **cropping system**
- **mint oil**
- **crop rotations**
- **intercropping**
- **relay cropping**

REFERENCES

1. Chaitra, S.M., Pushpa, T.N., Srikantaprasad, D., Hiremath, J.S. Biradar, I.B., Nandimath, S.T. Growth of menthol mint (*Mentha arvensis* L.) and sugarcane (*Saccharum officinarum* L.) as influenced by intercropping system. *The Pharmaceutical Journal*, 2021, *10*(7), 981–984.
2. Evans, J., Mcneill, A.M., Unkovich, M.J., Fettell, N.A., Heenan, D.P. Net nitrogen balances for cool-season grain legume intercropping and contributions to wheat nitrogen uptake: A review. *Australian Journal of Experimental Agriculture*, 2001, *41*, 347–359.
3. Gill, B.S., Gill, G.S., Salaria, A. Intercropping Japanese mint (Mentha arvensis L.) with maize, oilseed rape and onion during rabi season. *Journal Spices, Aromatic and Crops*, 2007, *16*(1), 2730.
4. Gowton, C.M., Arias, C.A., Carrillo, J. Intercropping with Pepper mint increases ground dwelling insect and pollinator abundance and decreases Drosophila suzukiiin fruit. *Frontiers in Sustainable Food Systems*, 2021, *5*.
5. Maia, J.T.L.S., Ismail, T., de, Souza, Júnior, I., Valadares, S.V., Ferraz, E.O., Martins, E.R., da Costa, C.A. Agronomic performance of lettuce and carrot, in monoculture and associated with medicinal plants, under organic systems. *Revista Brasileira de Ciências Agrárias*, 2009, *4*(3), 283–288.
6. Joshi, K., Singh, V., Kumar, A., Tiwari R. Effect on herbage and oil yield in different Mentha species intercropped with poplar. *Journal Pharmacognosy and Phytochemistry*, 2018, *1*, 1750–1754.
7. Khippal, A., Singh, S., Shokand, R., Singh, J., Sharma, A.K., Verma., R. Mechanized intercropping of legumes in spring planted sugarcane. *Vegetos*, 2016, *29*(1), 52–64.
8. Kothari, S.K., Singh, J.P., Singh, V. Intercropping of mint species in spring planted sugarcane under tarai conditions of Uttar Pradesh. *Journal of Science and Technology*, 1987, *4*, 1–6.
9. Kumar, S., Bahl, J.R., Bansal, R.P., Gupta, A.K, Singh, V., Sharma, S. High economic returns from companion and relay cropping of bread wheat and menthol mint in the winter-summer season in north Indian plains. *Industrial Crops and Products*, 2002, *15*(2), 103–114.
10. Kumar, R., Singh, J., Uppal, S.K. Intercropping of medicinal and high value crops in autumn sugarcane (*Saccharum* spp. hybrid) for higher productivity and profitability. *Indian Journal of Agronomy*, 2015, *60*(1), 61–64.
11. Machiani, M.A., Javanmard, A., Morshedloo, M.R., Maggi, F. Evaluation of competition, essential oil quality and quantity of peppermint intercropped with soybean. *Industrial Crops & Products,* 2018, *111*, 743754.
12. Machiani, M.A., Javanmard, A., Nasiri, Y., Morshedloo, M.R. Advantage of peppermint (*Mentha piperita* L.) and Faba Bean (Viciafaba L.). *Journal of Agricultural Science and Sustainable Production*, 2017, *27*(3), 45–62.
13. Mitchell, A.R. Intercropping in different CroppingPatterns. Peppermint relay intercropping with rye. *Journal of Sustainable Agriculture*, 1996, *7*(4), 103–110.
14. Randhawa, G.S., Sidhu, B.S., Mahey, R.K. Intercropping of mentha in sugarcane. *Indian Journal of Agronomy*, 1989, *34*(4), 4988–500

15. Telci, I., Kacar, O., Bayram, E., Arabac, O., Demirta, I., Ylmaz, G., Ozcan, I., Sönmez, C., Göksu, E. The effect of ecological conditions on yield and quality traits of selected peppermint (*Mentha piperita* L.) clones. *Industrial Crops and Products*, 2011, *34*(1), 1193–1197.
16. Zarekar, V.V. *Effect of Intercropping and Planting Methods on Growth, Yield and Quality of Sugarcane Under Lateritic Soil Condition.* MSc (Agriculture). Thesis, Dr. Balasheb Sawant Konkan Krishi Vidyapeeth, Dapoli, Maharashtra, India, 2016.

Index

R

For Product Safety Concerns and Information please contact our EU
representative GPSR@taylorandfrancis.com
Taylor & Francis Verlag GmbH, Kaufingerstraße 24, 80331 München, Germany

www.ingramcontent.com/pod-product-compliance
Ingram Content Group UK Ltd.
Pitfield, Milton Keynes, MK11 3LW, UK
UKHW022158190926
475702UK00003B/95
9781779641540